Teacher, Student, and Parent
One-Stop Internet Resources

Log on to
booki.msscience.com

ONLINE STUDY TOOLS

- Section Self-Check Quizzes
- Interactive Tutor
- Chapter Review Tests
- Standardized Test Practice
- Vocabulary PuzzleMaker

ONLINE RESEARCH

- WebQuest Projects
- Prescreened Web Links
- Career Links
- Internet Labs

INTERACTIVE ONLINE STUDENT EDITION

- Complete Interactive Student Edition available at mhln.com

FOR TEACHERS

- Teacher Bulletin Board
- Teaching Today—Professional Development

SAFETY SYMBOLS	HAZARD	EXAMPLES	PRECAUTION	REMEDY
DISPOSAL	Special disposal procedures need to be followed.	certain chemicals, living organisms	Do not dispose of these materials in the sink or trash can.	Dispose of wastes as directed by your teacher.
BIOLOGICAL	Organisms or other biological materials that might be harmful to humans	bacteria, fungi, blood, unpreserved tissues, plant materials	Avoid skin contact with these materials. Wear mask or gloves.	Notify your teacher if you suspect contact with material. Wash hands thoroughly.
EXTREME TEMPERATURE	Objects that can burn skin by being too cold or too hot	boiling liquids, hot plates, dry ice, liquid nitrogen	Use proper protection when handling.	Go to your teacher for first aid.
SHARP OBJECT	Use of tools or glassware that can easily puncture or slice skin	razor blades, pins, scalpels, pointed tools, dissecting probes, broken glass	Practice common-sense behavior and follow guidelines for use of the tool.	Go to your teacher for first aid.
FUME	Possible danger to respiratory tract from fumes	ammonia, acetone, nail polish remover, heated sulfur, moth balls	Make sure there is good ventilation. Never smell fumes directly. Wear a mask.	Leave foul area and notify your teacher immediately.
ELECTRICAL	Possible danger from electrical shock or burn	improper grounding, liquid spills, short circuits, exposed wires	Double-check setup with teacher. Check condition of wires and apparatus.	Do not attempt to fix electrical problems. Notify your teacher immediately.
IRRITANT	Substances that can irritate the skin or mucous membranes of the respiratory tract	pollen, moth balls, steel wool, fiberglass, potassium permanganate	Wear dust mask and gloves. Practice extra care when handling these materials.	Go to your teacher for first aid.
CHEMICAL	Chemicals can react with and destroy tissue and other materials	bleaches such as hydrogen peroxide; acids such as sulfuric acid, hydrochloric acid; bases such as ammonia, sodium hydroxide	Wear goggles, gloves, and an apron.	Immediately flush the affected area with water and notify your teacher.
TOXIC	Substance may be poisonous if touched, inhaled, or swallowed.	mercury, many metal compounds, iodine, poinsettia plant parts	Follow your teacher's instructions.	Always wash hands thoroughly after use. Go to your teacher for first aid.
FLAMMABLE	Flammable chemicals may be ignited by open flame, spark, or exposed heat.	alcohol, kerosene, potassium permanganate	Avoid open flames and heat when using flammable chemicals.	Notify your teacher immediately. Use fire safety equipment if applicable.
OPEN FLAME	Open flame in use, may cause fire.	hair, clothing, paper, synthetic materials	Tie back hair and loose clothing. Follow teacher's instruction on lighting and extinguishing flames.	Notify your teacher immediately. Use fire safety equipment if applicable.

 Eye Safety
Proper eye protection should be worn at all times by anyone performing or observing science activities.

 Clothing Protection
This symbol appears when substances could stain or burn clothing.

 Animal Safety
This symbol appears when safety of animals and students must be ensured.

 Handwashing
After the lab, wash hands with soap and water before removing goggles.

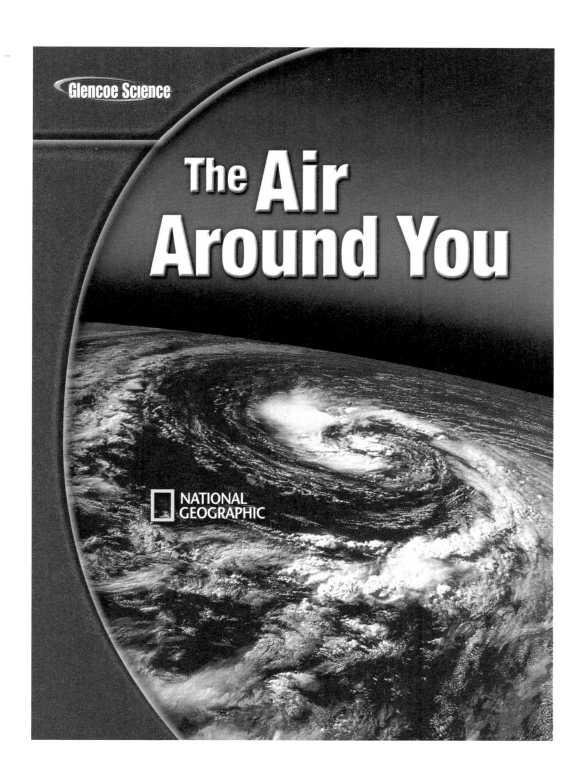

Glencoe Science

The Air Around You

NATIONAL GEOGRAPHIC

Mc Graw Hill **Glencoe**

New York, New York   Columbus, Ohio   Chicago, Illinois   Woodland Hills, California

Glencoe Science

The Air Around You

This satellite image shows Hurricane Bonnie, which struck North Carolina in 1998. The storm was nearly 400 miles wide, with the highest recorded wind gust at 104 mph. Overall damages were estimated in the $1.0 billion dollar range, and three deaths were attributed to the Category 3 storm.

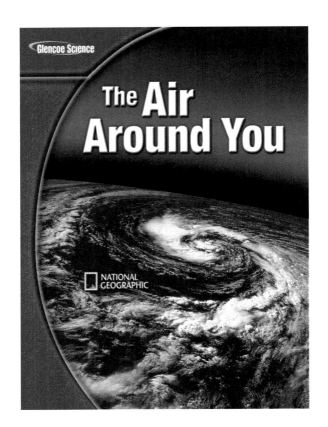

Glencoe Science

The **Air Around You**

NATIONAL GEOGRAPHIC

Glencoe

The *McGraw·Hill* Companies

Send all inquiries to:
Glencoe/McGraw-Hill
8787 Orion Place
Columbus, OH 43240-4027

ISBN: 978-0-07-877828-5
MHID: 0-07-877828-X

Printed in the United States of America.

1 2 3 4 5 6 7 8 9 10 027/043 09 08 07

Authors

NATIONAL GEOGRAPHIC
Education Division
Washington, D.C.

Susan Leach Snyder
Earth Science Teacher, Consultant
Jones Middle School
Upper Arlington, OH

Dinah Zike
Educational Consultant
Dinah-Might Activities, Inc.
San Antonio, TX

Series Consultants

CONTENT

William C. Keel, PhD
Department of Physics and
Astronomy
University of Alabama
Tuscaloosa, AL

MATH

Teri Willard, EdD
Mathematics Curriculum Writer
Belgrade, MT

READING

Carol A. Senf, PhD
School of Literature,
Communication, and Culture
Georgia Institute of Technology
Atlanta, GA

SAFETY

Aileen Duc, PhD
Science 8 Teacher
Hendrick Middle School, Plano ISD
Plano, TX

ACTIVITY TESTERS

Nerma Coats Henderson
Pickerington Lakeview Jr. High
School
Pickerington, OH

Mary Helen Mariscal-Cholka
William D. Slider Middle School
El Paso, TX

**Science Kit and Boreal
Laboratories**
Tonawanda, NY

Series Reviewers

Lois Burdette
Green Bank Elementary-Middle
School
Green Bank, WV

Marcia Chackan
Pine Crest School
Boca Raton, FL

Annette D'Urso Garcia
Kearney Middle School
Commerce City, CO

Nerma Coats Henderson
Pickerington Lakeview Jr. High
School
Pickerington, OH

Michael Mansour
Board Member
National Middle Level Science
Teacher's Association
John Page Middle School
Madison Heights, MI

Sharon Mitchell
William D. Slider Middle School
El Paso, TX

Mark Sailer
Pioneer Jr-Sr. High School
Royal Center, IN

Kate Ziegler
Durant Road Middle School
Raleigh, NC

HOW TO...

Use Your Science Book

Before You Read

- **Chapter Opener** Science is occurring all around you, and the opening photo of each chapter will preview the science you will be learning about. The **Chapter Preview** will give you an idea of what you will be learning about, and you can try the **Launch Lab** to help get your brain headed in the right direction. The **Foldables** exercise is a fun way to keep you organized.

- **Section Opener** Chapters are divided into two to four sections. The **As You Read** in the margin of the first page of each section will let you know what is most important in the section. It is divided into four parts. **What You'll Learn** will tell you the major topics you will be covering. **Why It's Important** will remind you why you are studying this in the first place! The **Review Vocabulary** word is a word you already know, either from your science studies or your prior knowledge. The **New Vocabulary** words are words that you need to learn to understand this section. These words will be in **boldfaced** print and highlighted in the section. Make a note to yourself to recognize these words as you are reading the section.

The **Air Around You**

As You Read

- **Headings** Each section has a title in large red letters, and is further divided into blue titles and small red titles at the beginnings of some paragraphs. To help you study, make an outline of the headings and subheadings.

- **Margins** In the margins of your text, you will find many helpful resources. The **Science Online** exercises and **Integrate** activities help you explore the topics you are studying. **MiniLabs** reinforce the science concepts you have learned.

- **Building Skills** You also will find an **Applying Math** or **Applying Science** activity in each chapter. This gives you extra practice using your new knowledge, and helps prepare you for standardized tests.

- **Student Resources** At the end of the book you will find **Student Resources** to help you throughout your studies. These include **Science, Technology,** and **Math Skill Handbooks,** an **English/Spanish Glossary,** and an **Index.** Also, use your **Foldables** as a resource. It will help you organize information, and review before a test.

- **In Class** Remember, you can always ask your teacher to explain anything you don't understand.

FOLDABLES™
Study Organizer

Science Vocabulary Make the following Foldable to help you understand the vocabulary terms in this chapter.

STEP 1 Fold a vertical sheet of notebook paper from side to side.

STEP 2 Cut along every third line of only the top layer to form tabs.

STEP 3 Label each tab with a vocabulary word from the chapter.

Build Vocabulary As you read the chapter, list the vocabulary words on the tabs. As you learn the definitions, write them under the tab for each vocabulary word.

Look For...

FOLDABLES™

At the beginning of every section.

In Lab

Working in the laboratory is one of the best ways to understand the concepts you are studying. Your book will be your guide through your laboratory experiences, and help you begin to think like a scientist. In it, you not only will find the steps necessary to follow the investigations, but you also will find helpful tips to make the most of your time.

- Each lab provides you with a **Real-World Question** to remind you that science is something you use every day, not just in class. This may lead to many more questions about how things happen in your world.

- Remember, experiments do not always produce the result you expect. Scientists have made many discoveries based on investigations with unexpected results. You can try the experiment again to make sure your results were accurate, or perhaps form a new hypothesis to test.

- Keeping a **Science Journal** is how scientists keep accurate records of observations and data. In your journal, you also can write any questions that may arise during your investigation. This is a great method of reminding yourself to find the answers later.

Look For...
- **Launch Labs** start every chapter.
- **MiniLabs** in the margin of each chapter.
- **Two Full-Period Labs** in every chapter.
- **EXTRA Try at Home Labs** at the end of your book.
- the **Web site** with laboratory demonstrations.

Before a Test

Admit it! You don't like to take tests! However, there *are* ways to review that make them less painful. Your book will help you be more successful taking tests if you use the resources provided to you.

- Review all of the **New Vocabulary** words and be sure you understand their definitions.

- Review the notes you've taken on your **Foldables,** in class, and in lab. Write down any question that you still need answered.

- Review the **Summaries** and **Self Check questions** at the end of each section.

- Study the concepts presented in the chapter by reading the **Study Guide** and answering the questions in the **Chapter Review.**

Look For...
- Reading Checks and caption questions throughout the text.
- the Summaries and Self Check questions at the end of each section.
- the Study Guide and Review at the end of each chapter.
- the Standardized Test Practice after each chapter.

Let's Get Started

To help you find the information you need quickly, use the Scavenger Hunt below to learn where things are located in Chapter 1.

1. What is the title of this chapter?

2. What will you learn in Section 1?

3. Sometimes you may ask, "Why am I learning this?" State a reason why the concepts from Section 2 are important.

4. What is the main topic presented in Section 2?

5. How many reading checks are in Section 1?

6. What is the Web address where you can find extra information?

7. What is the main heading above the sixth paragraph in Section 2?

8. There is an integration with another subject mentioned in one of the margins of the chapter. What subject is it?

9. List the new vocabulary words presented in Section 2.

10. List the safety symbols presented in the first Lab.

11. Where would you find a Self Check to be sure you understand the section?

12. Suppose you're doing the Self Check and you have a question about concept mapping. Where could you find help?

13. On what pages are the Chapter Study Guide and Chapter Review?

14. Look in the Table of Contents to find out on which page Section 2 of the chapter begins.

15. You complete the Chapter Review to study for your chapter test. Where could you find another quiz for more practice?

Teacher Advisory Board

The Teacher Advisory Board gave the editorial staff and design team feedback on the content and design of the Student Edition. They provided valuable input in the development of the 2008 edition of *Glencoe Science.*

John Gonzales
Challenger Middle School
Tucson, AZ

Rachel Shively
Aptakisic Jr. High School
Buffalo Grove, IL

Roger Pratt
Manistique High School
Manistique, MI

Kirtina Hile
Northmor Jr. High/High School
Galion, OH

Marie Renner
Diley Middle School
Pickerington, OH

Nelson Farrier
Hamlin Middle School
Springfield, OR

Jeff Remington
Palmyra Middle School
Palmyra, PA

Erin Peters
Williamsburg Middle School
Arlington, VA

Rubidel Peoples
Meacham Middle School
Fort Worth, TX

Kristi Ramsey
Navasota Jr. High School
Navasota, TX

Student Advisory Board

The Student Advisory Board gave the editorial staff and design team feedback on the design of the Student Edition. We thank these students for their hard work and creative suggestions in making the 2008 edition of *Glencoe Science* student friendly.

Jack Andrews
Reynoldsburg Jr. High School
Reynoldsburg, OH

Peter Arnold
Hastings Middle School
Upper Arlington, OH

Emily Barbe
Perry Middle School
Worthington, OH

Kirsty Bateman
Hilliard Heritage Middle School
Hilliard, OH

Andre Brown
Spanish Emersion Academy
Columbus, OH

Chris Dundon
Heritage Middle School
Westerville, OH

Ryan Manafee
Monroe Middle School
Columbus, OH

Addison Owen
Davis Middle School
Dublin, OH

Teriana Patrick
Eastmoor Middle School
Columbus, OH

Ashley Ruz
Karrar Middle School
Dublin, OH

The Glencoe middle school science Student Advisory Board taking a timeout at COSI, a science museum in Columbus, Ohio.

Contents

In each chapter, look for these opportunities for review and assessment:
- Reading Checks
- Caption Questions
- Section Review
- Chapter Study Guide
- Chapter Review
- Standardized Test Practice
- Online practice at booki.msscience.com

Get Ready to Read Strategies

Student Resources

Cross-Curricular Readings/Labs

DVD available as a video lab

NATIONAL GEOGRAPHIC VISUALIZING

TIME SCIENCE AND Society

TIME SCIENCE AND HISTORY

Oops! Accidents in SCIENCE

Science and Language Arts

Launch LAB

Mini LAB

Mini LAB Try at Home

One-Page Labs

Two-Page Labs

Design Your Own Labs

Model and Invent Labs

Content Details

Content Details

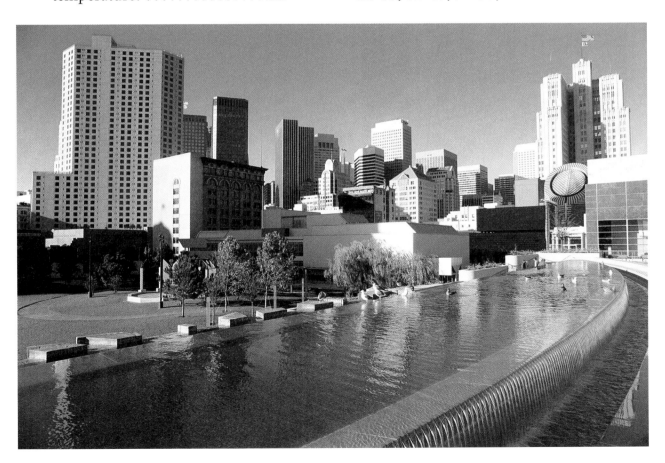

Storm Scientists

Hurricanes are among nature's most destructive forces. These storms, which have lasting wind speeds of at least 120 km/h, can flatten trees, destroy houses, and kill people. Satellite images of hurricanes help scientists estimate when, where, and with how much force a storm will strike. However, researchers sometimes need detailed information about the internal structure of a hurricane that satellites can't provide. To collect such information, daring scientists fly airplanes where no other travelers dare—directly into the strongest winds of a hurricane.

To measure the fury of a hurricane, researchers must punch through the eye wall—a swirling wall of clouds with high winds surrounding the eye. Sometimes the clouds are so thick in the eye wall that the crew can't see the airplane's wings.

Figure 1 The roof was ripped off this home in Hawaii by the powerful winds of a hurricane.

Figure 2 Scientists use aircraft fitted with high-tech measuring devices to fly into hurricanes and collect data.

Gathering Information

Researchers who fly into hurricanes are blinded by rain. Powerful winds can send the aircraft plummeting to Earth. Despite these dangers, scientists continue to make measurements that require flying into hurricanes.

The NOAA (National Oceanic and Atmospheric Administration) maintains special planes fitted with wind probes and other devices to collect data from hurricanes. These planes fly at maximum and minimum elevations of 6,000 m and 450 m. The low-altitude flying is particularly dangerous because there is little room for recovery if a plane loses control. While flying through hurricanes, scientists sometimes release dropsondes, which are small devices that parachute down through a storm taking measurements such as temperature, pressure, wind direction, and humidity.

These data are used in computer models that predict how intense a hurricane is and where it might reach land. The models, in turn, are used to issue watches, warnings, and forecasts to minimize destruction of property and loss of life.

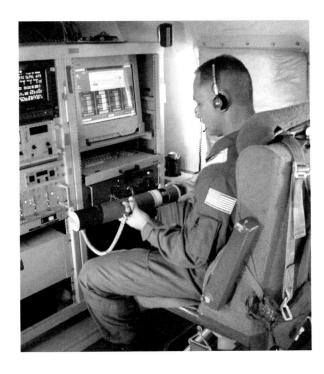

Figure 3 Dropsondes released from aircraft collect information as they fall through a hurricane.

Figure 4 Hurricane Georges wreaked havoc in Key West, Florida, in 1998.

SCIENCE

The Study of Weather

Meteorologists are scientists who study weather and make predictions. Meteorologists make weather forecasts using data collected from measurements and observations. Forecasting the strength and movement of hurricanes becomes more accurate if data are collected from many regions.

Even when flying into the eye of a hurricane, researchers must use scientific methods and make accurate observations to make predictions about a storm's strength and direction. Many of these data are gathered by performing a variety of sophisticated investigations, or experiments.

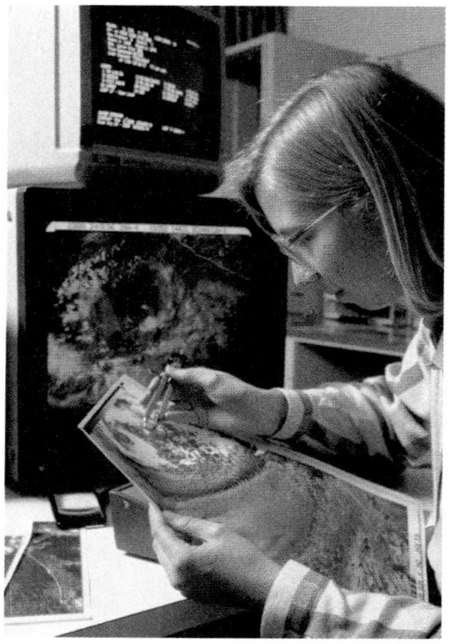

Experimentation

Scientists try to answer questions by performing tests, called experiments, and recording the results. Experiments must be carefully planned in order to ensure the accuracy of the results. Scientists begin by making educated guesses, called hypotheses, about what the results of an experiment might be. Hurricane researchers, for example, hypothesize about what specific data would be most useful for predicting the path of a hurricane.

Variables, Constants, and Controls

When scientists conduct experiments, they try to make sure that only one factor affects the results of the experiment. The factor that is changed in the experiment is called the independent variable. The dependent variable is what is measured or observed in the experiment. Many experiments use a control—a sample that is treated like all the others except that the independent variable isn't applied. Conditions that stay the same in an experiment are called constants.

Constants in hurricane research include using the same methods and devices to measure the air pressure and the wind strength. However, it's impossible to isolate one independent variable or to use a control. In the case of hurricane research, many observations are obtained and analyzed to compensate for this before conclusions are drawn. In addition, scientists who study hurricanes use computer models to create virtual storms in which they are able to manipulate one independent variable at a time.

Figure 5 Meteorologists study Earth's atmosphere in order to predict hurricanes and other, less extreme weather changes.

Interpreting Data

The observations and measurements that a scientist makes in an experiment are called data. In addition to obtaining images of hurricanes using satellites, scientists who fly aircraft through hurricanes collect many types of data, such as temperature, humidity, wind speed, and wind direction. Data must be carefully organized and studied before questions can be answered or problems can be solved.

Drawing Conclusions

An important step in any scientific method is to draw a conclusion based on results and observations. A conclusion summarizes what researchers have learned from an experiment.

Timely and accurate conclusions are important in hurricane research. Those conclusions lead to predictions of when and where a hurricane might strike and with what intensity. The predictions then must be communicated to the public. The people who make predictions about where hurricanes might reach land must consider many factors. Safety is the most important concern. However, forecasters also must be careful not to make premature predictions. Hurricane forecasts cause people to prepare their property for the storm, and evacuate the region. These things cost money and can impact local economies. These issues can weigh heavily on the minds of those who issue hurricane watches and warnings.

Figure 6 Scientists such as Max Mayfield of the National Hurricane Center in Miami, Florida, are responsible for issuing hurricane watches and warnings.

Meteorologists use sophisticated equipment to predict the paths of hurricanes, but the paths of smaller storms often can be predicted reliably with weather maps, barometers, and other common equipment. Design a weather station that could be built at your school to predict when storms might reach your area. Describe how you would use data collected from the station.

Atmosphere

The BIG Idea

Earth's atmosphere helps regulate the absorption and distribution of energy received from the Sun.

SECTION 1
Earth's Atmosphere
Main Idea Earth's atmosphere is a thin layer of air that forms a protective covering around the planet.

SECTION 2
Energy Transfer in the Atmosphere
Main Idea Earth's atmosphere helps control how much of the Sun's radiation is absorbed or lost to space.

SECTION 3
Air Movement
Main Idea Uneven heating of Earth's surface leads to a change in pressure that causes air to move.

Fresh mountain air?

On top of Mt. Everest the air is a bit thin. Without breathing equipment, an average person quickly would become dizzy, then unconscious, and eventually would die. In this chapter you'll learn what makes the atmosphere at high altitudes different from the atmosphere we are used to.

Science Journal Write a short article describing how you might prepare to climb Mt. Everest.

Start-Up Activities

Observe Air Pressure

The air around you is made of billions of molecules. These molecules are constantly moving in all directions and bouncing into every object in the room, including you. Air pressure is the result of the billions of collisions of molecules into these objects. Because you usually do not feel molecules in air hitting you, do the lab below to see the effect of air pressure.

1. Cut out a square of cardboard about 10 cm from the side of a cereal box.

2. Fill a glass to the brim with water.

3. Hold the cardboard firmly over the top of the glass, covering the water, and invert the glass.

4. Slowly remove your hand holding the cardboard in place and observe.

5. **Think Critically** Write a paragraph in your Science Journal describing what happened to the cardboard when you inverted the glass and removed your hand. How does air pressure explain what happened?

Earth's Atmospheric Layers
Make the following Foldable to help you visualize the five layers of Earth's atmosphere.

STEP 1 Collect 3 sheets of paper and layer them about 1.25 cm apart vertically. Keep the edges level.

STEP 2 Fold up the bottom edges of the paper to form 6 equal tabs.

STEP 3 Fold the paper and crease well to hold the tabs in place. Staple along the fold. Label each tab.

Exosphere
Thermosphere
Mesosphere
Stratosphere
Troposphere
Earth's Atmosphere

Find Main Ideas Label the tabs *Earth's Atmosphere, Troposphere, Stratosphere, Mesosphere, Thermosphere,* and *Exosphere* from bottom to top as shown. As you read the chapter, write information about each layer of Earth's atmosphere under the appropriate tab.

Science Online

Preview this chapter's content and activities at
booki.msscience.com

Identify the Main Idea

1 Learn It! Main ideas are the most important ideas in a paragraph, section, or chapter. Supporting details are facts or examples that explain the main idea. Understanding the main idea allows you to grasp the whole picture.

2 Picture It! Read the following paragraph. Draw a graphic organizer like the one below to show the main idea and supporting details.

> In addition to gases, Earth's atmosphere contains small, solid particles such as dust, salt, and pollen. Dust particles get into the atmosphere when wind picks them up off the ground and carries them along. Salt is picked up from ocean spray. Plants give off pollen that becomes mixed throughout part of the atmosphere.
>
> —*from page 9*

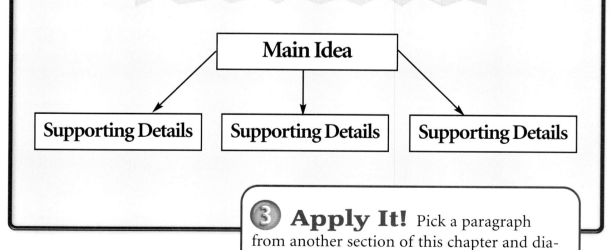

3 Apply It! Pick a paragraph from another section of this chapter and diagram the main ideas as you did above.

Reading Tip

The main idea is often the first sentence in a paragraph but not always.

Target Your Reading

Use this to focus on the main ideas as you read the chapter.

1 **Before you read** the chapter, respond to the statements below on your worksheet or on a numbered sheet of paper.

- Write an **A** if you **agree** with the statement.
- Write a **D** if you **disagree** with the statement.

2 **After you read** the chapter, look back to this page to see if you've changed your mind about any of the statements.

- If any of your answers changed, explain why.
- Change any false statements into true statements.
- Use your revised statements as a study guide.

Science Online
Print out a worksheet of this page at
booki.msscience.com

Before You Read A or D		Statement	After You Read A or D
	1	Earth's atmosphere is mostly oxygen.	
	2	Air pressure is greater near Earth's surface and decreases higher in the atmosphere.	
	3	The ozone layer absorbs most of the harmful infrared radiation that enters the atmosphere.	
	4	Conduction is the transfer of heat by the flow of material.	
	5	In the atmosphere, cold, dense air sinks, causing hot, less dense air to rise.	
	6	Wind is the movement of air from an area of lower pressure to an area of higher pressure.	
	7	Earth's surface is heated evenly by the Sun.	
	8	Earth's rotation affects the direction in which air and water move.	
	9	Jet streams are legally defined zones in the atmosphere where only jets are allowed to travel.	

Earth's Atmosphere

Importance of the Atmosphere

Earth's **atmosphere,** shown in **Figure 1,** is a thin layer of air that forms a protective covering around the planet. If Earth had no atmosphere, days would be extremely hot and nights would be extremely cold. Earth's atmosphere maintains a balance between the amount of heat absorbed from the Sun and the amount of heat that escapes back into space. It also protects life-forms from some of the Sun's harmful rays.

Makeup of the Atmosphere

Earth's atmosphere is a mixture of gases, solids, and liquids that surrounds the planet. It extends from Earth's surface to outer space. The atmosphere is much different today from what it was when Earth was young.

Earth's early atmosphere, produced by erupting volcanoes, contained nitrogen and carbon dioxide, but little oxygen. Then, more than 2 billon years ago, Earth's early organisms released oxygen into the atmosphere as they made food with the aid of sunlight. These early organisms, however, were limited to layers of ocean water deep enough to be shielded from the Sun's harmful rays, yet close enough to the surface to receive sunlight. Eventually, a layer rich in ozone (O_3) that protects Earth from the Sun's harmful rays formed in the upper atmosphere. This protective layer eventually allowed green plants to flourish all over Earth, releasing even more oxygen. Today, a variety of life forms, including you, depends on a certain amount of oxygen in Earth's atmosphere.

Figure 1 Earth's atmosphere, as viewed from space, is a thin layer of gases. The atmosphere keeps Earth's temperature in a range that can support life.

Gases in the Atmosphere Today's atmosphere is a mixture of the gases shown in **Figure 2.** Nitrogen is the most abundant gas, making up 78 percent of the atmosphere. Oxygen actually makes up only 21 percent of Earth's atmosphere. As much as four percent of the atmosphere is water vapor. Other gases that make up Earth's atmosphere include argon and carbon dioxide.

The composition of the atmosphere is changing in small but important ways. For example, car exhaust emits gases into the air. These pollutants mix with oxygen and other chemicals in the presence of sunlight and form a brown haze called smog. Humans burn fuel for energy. As fuel is burned, carbon dioxide is released as a by-product into Earth's atmosphere. Increasing energy use may increase the amount of carbon dioxide in the atmosphere.

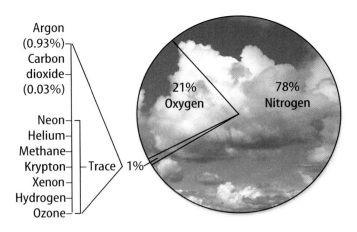

Figure 2 This circle graph shows the percentages of the gases, excluding water vapor, that make up Earth's atmosphere.
Determine *Approximately what fraction of Earth's atmosphere is oxygen?*

Solids and Liquids in Earth's Atmosphere In addition to gases, Earth's atmosphere contains small, solid particles such as dust, salt, and pollen. Dust particles get into the atmosphere when wind picks them up off the ground and carries them along. Salt is picked up from ocean spray. Plants give off pollen that becomes mixed throughout part of the atmosphere.

The atmosphere also contains small liquid droplets other than water droplets in clouds. The atmosphere constantly moves these liquid droplets and solids from one region to another. For example, the atmosphere above you may contain liquid droplets and solids from an erupting volcano thousands of kilometers from your home, as illustrated in **Figure 3.**

Figure 3 Solids and liquids can travel large distances in Earth's atmosphere, affecting regions far from their source.

On June 12, 1991, Mount Pinatubo in the Philippines erupted, causing liquid droplets to form in Earth's atmosphere.

Droplets of sulfuric acid from volcanoes can produce spectacular sunrises.

Layers of the Atmosphere

What would happen if you left a glass of chocolate milk on the kitchen counter for a while? Eventually, you would see a lower layer with more chocolate separating from upper layers with less chocolate. Like a glass of chocolate milk, Earth's atmosphere has layers. There are five layers in Earth's atmosphere, each with its own properties, as shown in **Figure 4.** The lower layers include the troposphere and stratosphere. The upper atmospheric layers are the mesosphere, thermosphere, and exosphere. The troposphere and stratosphere contain most of the air.

Lower Layers of the Atmosphere You study, eat, sleep, and play in the **troposphere** which is the lowest of Earth's atmospheric layers. It contains 99 percent of the water vapor and 75 percent of the atmospheric gases. Rain, snow, and clouds occur in the troposphere, which extends up to about 10 km.

The stratosphere, the layer directly above the troposphere, extends from 10 km above Earth's surface to about 50 km. As **Figure 4** shows, a portion of the stratosphere contains higher levels of a gas called ozone. Each molecule of ozone is made up of three oxygen atoms bonded together. Later in this section you will learn how ozone protects Earth from the Sun's harmful rays.

Topic: Earth's Atmospheric Layers

Visit booki.msscience.com for Web links to information about layers of Earth's atmosphere.

Activity Locate data on recent ozone layer depletion. Graph your data.

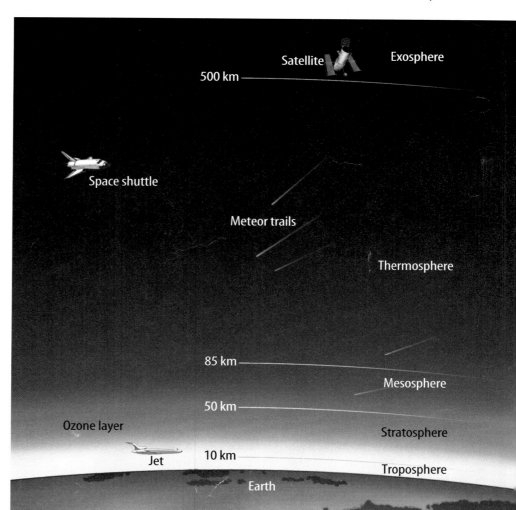

Figure 4 Earth's atmosphere is divided into five layers.
Describe *the layer of the atmosphere in which you live.*

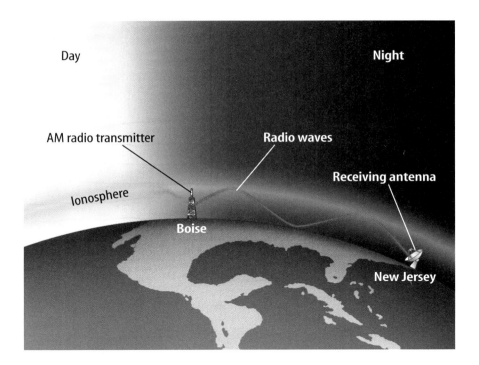

Day

Night

AM radio transmitter

Radio waves

Receiving antenna

Ionosphere

Boise

New Jersey

Figure 5 During the day, the ionosphere absorbs radio transmissions. This prevents you from hearing distant radio stations. At night, the ionosphere reflects radio waves. The reflected waves can travel to distant cities.
Describe *what causes the ionosphere to change between day and night.*

Upper Layers of the Atmosphere Beyond the stratosphere are the mesosphere, thermosphere, and exosphere. The mesosphere extends from the top of the stratosphere to about 85 km above Earth. If you've ever seen a shooting star, you might have witnessed a meteor in the mesosphere.

The thermosphere is named for its high temperatures. This is the thickest atmospheric layer and is found between 85 km and 500 km above Earth's surface.

Within the mesosphere and thermosphere is a layer of electrically charged particles called the **ionosphere** (i AH nuh sfihr). If you live in New Jersey and listen to the radio at night, you might pick up a station from Boise, Idaho. The ionosphere allows radio waves to travel across the country to another city, as shown in **Figure 5.** During the day, energy from the Sun interacts with the particles in the ionosphere, causing them to absorb AM radio frequencies. At night, without solar energy, AM radio transmissions reflect off the ionosphere, allowing radio transmissions to be received at greater distances.

The space shuttle in **Figure 6** orbits Earth in the exosphere. In contrast to the troposphere, the layer you live in, the exosphere has so few molecules that the wings of the shuttle are useless. In the exosphere, the spacecraft relies on bursts from small rocket thrusters to move around. Beyond the exosphere is outer space.

 **Reading Check** *How does the space shuttle maneuver in the exosphere?*

Figure 6 Wings help move aircraft in lower layers of the atmosphere. The space shuttle can't use its wings to maneuver in the exosphere because so few molecules are present.

Atmospheric Pressure

Imagine you're a football player running with the ball. Six players tackle you and pile one on top of the other. Who feels the weight more—you or the player on top? Like molecules anywhere else, atmospheric gases have mass. Atmospheric gases extend hundreds of kilometers above Earth's surface. As Earth's gravity pulls the gases toward its surface, the weight of these gases presses down on the air below. As a result, the molecules nearer Earth's surface are closer together. This dense air exerts more force than the less dense air near the top of the atmosphere. Force exerted on an area is known as pressure.

Like the pile of football players, air pressure is greater near Earth's surface and decreases higher in the atmosphere, as shown in **Figure 7.** People find it difficult to breathe in high mountains because fewer molecules of air exist there. Jets that fly in the stratosphere must maintain pressurized cabins so that people can breathe.

Figure 7 Air pressure decreases as you go higher in Earth's atmosphere.

 **Reading Check** *Where is air pressure greater—in the exosphere or in the troposphere?*

Applying Science

How does altitude affect air pressure?

Atmospheric gases extend hundreds of kilometers above Earth's surface, but the molecules that make up these gases are fewer and fewer in number as you go higher. This means that air pressure decreases with altitude.

Identifying the Problem

The graph on the right shows these changes in air pressure. Note that altitude on the graph goes up only to 50 km. The troposphere and the stratosphere are represented on the graph, but other layers of the atmosphere are not. By examining the graph, can you understand the relationship between altitude and pressure?

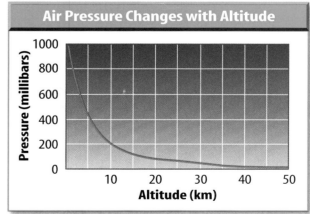

Solving the Problem
1. Estimate the air pressure at an altitude of 5 km.
2. Does air pressure change more quickly at higher altitudes or at lower altitudes?

Temperature in Atmospheric Layers

The Sun is the source of most of the energy on Earth. Before it reaches Earth's surface, energy from the Sun must pass through the atmosphere. Because some layers contain gases that easily absorb the Sun's energy while other layers do not, the various layers have different temperatures, illustrated by the red line in **Figure 8.**

Molecules that make up air in the troposphere are warmed mostly by heat from Earth's surface. The Sun warms Earth's surface, which then warms the air above it. When you climb a mountain, the air at the top is usually cooler than the air at the bottom. Every kilometer you climb, the air temperature decreases about 6.5°C.

Molecules of ozone in the stratosphere absorb some of the Sun's energy. Energy absorbed by ozone molecules raises the temperature. Because more ozone molecules are in the upper portion of the stratosphere, the temperature in this layer rises with increasing altitude.

Like the troposphere, the temperature in the mesosphere decreases with altitude. The thermosphere and exosphere are the first layers to receive the Sun's rays. Few molecules are in these layers, but each molecule has a great deal of energy. Temperatures here are high.

Mini LAB

Determining if Air Has Mass

Procedure 🚫 🧤
1. On a **pan balance**, find the mass of an **inflatable ball** that is completely deflated.
2. Hypothesize about the change in the mass of the ball when it is inflated.
3. Inflate the ball to its maximum recommended inflation pressure.
4. Determine the mass of the fully inflated ball.

Analysis
1. What change occurs in the mass of the ball when it is inflated?
2. Infer from your data whether air has mass.

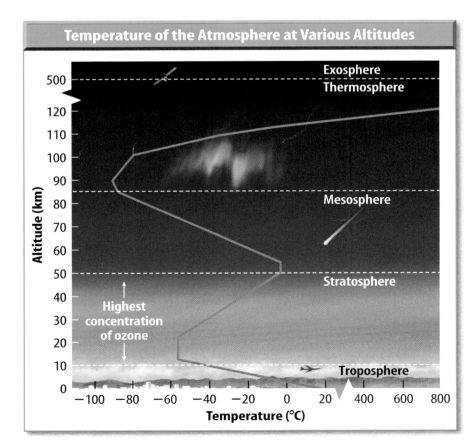

Temperature of the Atmosphere at Various Altitudes

Altitude (km) — Exosphere, Thermosphere, Mesosphere, Stratosphere, Troposphere. Highest concentration of ozone. Temperature (°C)

Figure 8 The division of the atmosphere into layers is based mainly on differences in temperature.
Determine *Does the temperature increase or decrease with altitude in the mesosphere?*

NEED TO KNOW

Effects of UV Light on Algae Algae are organisms that use sunlight to make their own food. This process releases oxygen to Earth's atmosphere. Some scientists suggest that growth is reduced when algae are exposed to ultraviolet radiation. Infer what might happen to the oxygen level of the atmosphere if increased ultraviolet radiation damages some algae.

Figure 9 Chlorofluorocarbon (CFC) molecules once were used in refrigerators and air conditioners. Each CFC molecule has three chlorine atoms. One atom of chlorine can destroy approximately 100,000 ozone molecules.

The Ozone Layer

Within the stratosphere, about 19 km to 48 km above your head, lies an atmospheric layer called the **ozone layer.** Ozone is made of oxygen. Although you cannot see the ozone layer, your life depends on it.

The oxygen you breathe has two atoms per molecule, but an ozone molecule is made up of three oxygen atoms bound together. The ozone layer contains a high concentration of ozone and shields you from the Sun's harmful energy. Ozone absorbs most of the ultraviolet radiation that enters the atmosphere. **Ultraviolet radiation** is one of the many types of energy that come to Earth from the Sun. Too much exposure to ultraviolet radiation can damage your skin and cause cancer.

CFCs Evidence exists that some air pollutants are destroying the ozone layer. Blame has fallen on **chlorofluorocarbons** (CFCs), chemical compounds used in some refrigerators, air conditioners, and aerosol sprays, and in the production of some foam packaging. CFCs can enter the atmosphere if these appliances leak or if they and other products containing CFCs are improperly discarded.

Recall that an ozone molecule is made of three oxygen atoms bonded together. Chlorofluorocarbon molecules, shown in **Figure 9,** destroy ozone. When a chlorine atom from a chlorofluorocarbon molecule comes near a molecule of ozone, the ozone molecule breaks apart. One of the oxygen atoms combines with the chlorine atom, and the rest form a regular, two-atom molecule. These compounds don't absorb ultraviolet radiation the way ozone can. In addition, the original chlorine atom can continue to break apart thousands of ozone molecules. The result is that more ultraviolet radiation reaches Earth's surface.

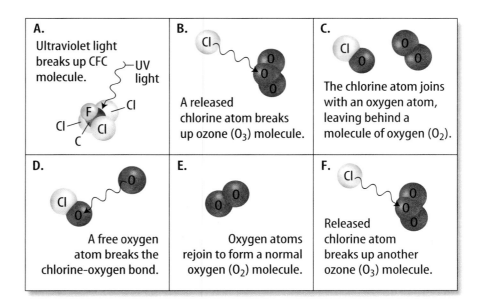

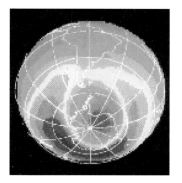
October 1980

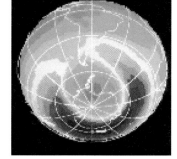
October 1988

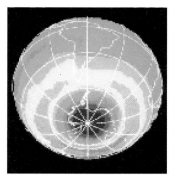
October 1990

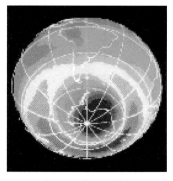
September 1999

INTEGRATE Health

The Ozone Hole The destruction of ozone molecules by CFCs seems to cause a seasonal reduction in ozone over Antarctica called the ozone hole. Every year beginning in late August or early September the amount of ozone in the atmosphere over Antarctica begins to decrease. By October, the ozone concentration reaches its lowest values and then begins to increase again. By December, the ozone hole disappears. **Figure 10** shows how the ozone hole over Antarctica has changed. In the mid-1990s, many governments banned the production and use of CFCs. Since then, the concentration of CFCs in the atmosphere has started to decrease.

Figure 10 These images of Antarctica were produced using data from a NASA satellite. The lowest values of ozone concentration are shown in dark blue and purple. These data show that the size of the seasonal ozone hole over Antarctica has grown larger over time.

section 1 review

Summary

Layers of the Atmosphere

- The atmosphere is a mixture of gases, solids, and liquids.
- The atmosphere has five layers—troposphere, stratosphere, mesosphere, thermosphere, and exosphere.
- The ionosphere is made up of electrically charged particles.

Atmospheric Pressure and Temperature

- Atmospheric pressure decreases with distance from Earth.
- Because some layers absorb the Sun's energy more easily than others, the various layers have different temperatures.

Ozone Layer

- The ozone layer absorbs most UV light.
- Chlorofluorocarbons (CFCs) break down the ozone layer.

Self Check

1. **Describe** How did oxygen come to make up 21 percent of Earth's present atmosphere?
2. **Infer** While hiking in the mountains, you notice that it is harder to breathe as you climb higher. Explain.
3. **State** some effects of a thinning ozone layer.
4. **Think Critically** Explain why, during the day, the radio only receives AM stations from a nearby city, while at night, you're able to hear a distant city's stations.

Applying Skills

5. **Interpret Scientific Illustrations** Using **Figure 2,** determine the total percentage of nitrogen and oxygen in the atmosphere. What is the total percentage of argon and carbon dioxide?
6. **Communicate** The names of the atmospheric layers end with the suffix *-sphere,* a word that means "ball." Find out what *tropo-, meso-, thermo-,* and *exo-* mean. Write their meanings in your Science Journal and explain if the layers are appropriately named.

Evaluating Sunscreens

Without protection, sun exposure can damage your health. Sunscreens protect your skin from UV radiation. In this lab, you will draw inferences using different sunscreen labels.

◉ Real-World Question

How effective are various brands of sunscreens?

Goals
- ■ **Draw inferences** based on labels on sunscreen brands.
- ■ **Compare** the effectiveness of different sunscreen brands for protection against the Sun.
- ■ **Compare** the cost of several sunscreen brands.

Materials
variety of sunscreens of different brand names

Safety Precautions

◉ Procedure

1. Make a data table in your Science Journal using the following headings: *Brand Name, SPF, Cost per Milliliter,* and *Misleading Terms.*

2. The Sun Protection Factor (SPF) tells you how long the sunscreen will protect you. For example, an SPF of 4 allows you to stay in the Sun four times longer than if you did not use sunscreen. Record the SPF of each sunscreen on your data table.

3. **Calculate** the cost per milliliter of each sunscreen brand.

4. Government guidelines say that terms like *sunblock* and *waterproof* are misleading because sunscreens can't block the Sun's rays, and they do wash off in water. List misleading terms in your data table for each brand.

Sunscreen Assessment			
Brand Name			
SPF			
Cost per Milliliter		Do not write in this book.	
Misleading Terms			

◉ Conclude and Apply

1. **Explain** why you need to use sunscreen.

2. **Evaluate** A minimum of SPF 15 is considered adequate protection for a sunscreen. An SPF greater than 30 is considered by government guidelines to be misleading because sunscreens wash or wear off. Evaluate the SPF of each sunscreen brand.

3. **Discuss** Considering the cost and effectiveness of all the sunscreen brands, discuss which you consider to be the best buy.

ℂommunicating Your Data

Create a poster on the proper use of sunscreens, and provide guidelines for selecting the safest product.

Energy Transfer in the Atmosphere

Energy from the Sun

The Sun provides most of Earth's energy. This energy drives winds and ocean currents and allows plants to grow and produce food, providing nutrition for many animals. When Earth receives energy from the Sun, three different things can happen to that energy, as shown in **Figure 11.** Some energy is reflected back into space by clouds, particles, and Earth's surface. Some is absorbed by the atmosphere or by land and water on Earth's surface.

Heat

Heat is energy that flows from an object with a higher temperature to an object with a lower temperature. Energy from the Sun reaches Earth's surface and heats it. Heat then is transferred through the atmosphere in three ways—radiation, conduction, and convection, as shown in **Figure 12.**

***What* You'll Learn**

- **Describe** what happens to the energy Earth receives from the Sun.
- **Compare and contrast** radiation, conduction, and convection.
- **Explain** the water cycle and its effect on weather patterns and climate.

***Why* It's Important**

The Sun provides energy to Earth's atmosphere, allowing life to exist.

Review Vocabulary
evaporation: when a liquid changes to a gas at a temperature below the liquid's boiling point

New Vocabulary
- radiation
- conduction
- convection
- hydrosphere
- condensation

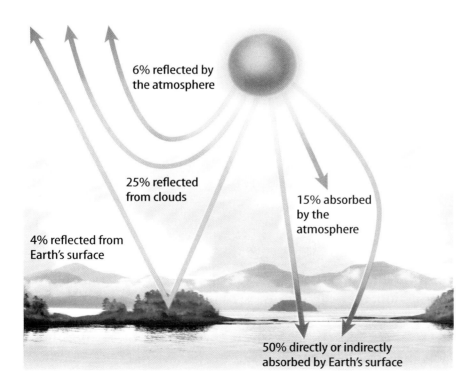

6% reflected by the atmosphere

25% reflected from clouds

4% reflected from Earth's surface

15% absorbed by the atmosphere

50% directly or indirectly absorbed by Earth's surface

Figure 11 The Sun is the source of energy for Earth's atmosphere. Thirty-five percent of incoming solar radiation is reflected back into space.
Infer *how much is absorbed by Earth's surface and atmosphere.*

Radiation warms the surface.

The air near Earth's surface is heated by conduction.

Cooler air pushes warm air upward, creating a convection current.

Figure 12 Heat is transferred within Earth's atmosphere by radiation, conduction, and convection.

Specific Heat Specific heat is the amount of heat required to raise the temperature of one kilogram of a substance one degree Celsius. Substances with high specific heat absorb a lot of heat for a small increase in temperature. Land warms faster than water does. Infer whether soil or water has a higher specific heat value.

Radiation Sitting on the beach, you feel the Sun's warmth on your face. How can you feel the Sun's heat even though you aren't in direct contact with it? Energy from the Sun reaches Earth in the form of radiant energy, or radiation. **Radiation** is energy that is transferred in the form of rays or waves. Earth radiates some of the energy it absorbs from the Sun back toward space. Radiant energy from the Sun warms your face.

✔ **Reading Check** *How does the Sun warm your skin?*

Conduction If you walk barefoot on a hot beach, your feet heat up because of conduction. **Conduction** is the transfer of energy that occurs when molecules bump into one another. Molecules are always in motion, but molecules in warmer objects move faster than molecules in cooler objects. When objects are in contact, energy is transferred from warmer objects to cooler objects.

Radiation from the Sun heated the beach sand, but direct contact with the sand warmed your feet. In a similar way, Earth's surface conducts energy directly to the atmosphere. As air moves over warm land or water, molecules in air are heated by direct contact.

Convection After the atmosphere is warmed by radiation or conduction, the heat is transferred by a third process called convection. **Convection** is the transfer of heat by the flow of material. Convection circulates heat throughout the atmosphere. How does this happen?

When air is warmed, the molecules in it move apart and the air becomes less dense. Air pressure decreases because fewer molecules are in the same space. In cold air, molecules move closer together. The air becomes more dense and air pressure increases. Cooler, denser air sinks while warmer, less dense air rises, forming a convection current. As **Figure 12** shows, radiation, conduction, and convection together distribute the Sun's heat throughout Earth's atmosphere.

The Water Cycle

Hydrosphere is a term that describes all the waters of Earth. The constant cycling of water within the atmosphere and the hydrosphere, as shown in **Figure 13,** plays an important role in determining weather patterns and climate types.

Energy from the Sun causes water to change from a liquid to a gas by a process called evaporation. Water that evaporates from lakes, streams, and oceans enters Earth's atmosphere. If water vapor in the atmosphere cools enough, it changes back into a liquid. This process of water vapor changing to a liquid is called **condensation.**

Clouds form when condensation occurs high in the atmosphere. Clouds are made up of tiny water droplets that can collide to form larger drops. As the drops grow, they fall to Earth as precipitation. This completes the water cycle within the hydrosphere. Classification of world climates is commonly based on annual and monthly averages of temperature and precipitation that are strongly affected by the water cycle.

Mini LAB

Modeling Heat Transfer

Procedure
1. Cover the outside of an empty **soup can,** with **black construction paper.**
2. Fill the can with **cold water** and feel it with your fingers.
3. Place the can in sunlight for 1 h, then pour the water over your fingers.

Analysis
1. Does the water in the can feel warmer or cooler after placing the can in sunlight?
2. What types of heat transfer did you model?

Try at Home

Figure 13 In the water cycle, water moves from Earth to the atmosphere and back to Earth again.

Precipitation

Condensation

Evaporation

Runoff

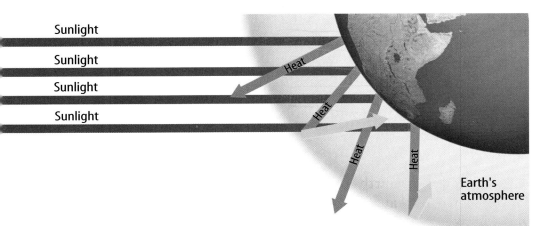

Sunlight

Sunlight

Sunlight

Sunlight

Heat

Heat

Heat

Heat

Earth's
atmosphere

Figure 14 Earth's atmosphere creates a delicate balance between energy received and energy lost. **Infer** What could happen if the balance is tipped toward receiving more energy than it does now?

Earth's Atmosphere is Unique

On Earth, radiation from the Sun can be reflected into space, absorbed by the atmosphere, or absorbed by land and water. Once it is absorbed, heat can be transferred by radiation, conduction, or convection. Earth's atmosphere, shown in **Figure 14,** helps control how much of the Sun's radiation is absorbed or lost.

✔ Reading Check *What helps control how much of the Sun's radiation is absorbed on Earth?*

Why doesn't life exist on Mars or Venus? Mars is a cold, lifeless world because its atmosphere is too thin to support life or to hold much of the Sun's heat. Temperatures on the surface of Mars range from 35°C to −170°C. On the other hand, gases in Venus's dense atmosphere trap heat coming from the Sun. The temperature on the surface of Venus is 470°C. Living things would burn instantly if they were placed on Venus's surface. Life on Earth exists because the atmosphere holds just the right amount of the Sun's energy.

section ② review

Summary

Energy From the Sun

- The Sun's radiation is either absorbed or reflected by Earth.
- Heat is transferred by radiation (waves), conduction (contact), or convection (flow).

The Water Cycle

- The water cycle affects climate.
- Water moves between the hydrosphere and the atmosphere through a continual process of evaporation and condensation.

Earth's Atmosphere is Unique

- Earth's atmosphere controls the amount of solar radiation that reaches Earth's surface.

Self Check

1. **State** how the Sun transfers energy to Earth.
2. **Contrast** the atmospheres of Earth and Mars.
3. **Describe** briefly the steps included in the water cycle.
4. **Explain** how the water cycle is related to weather patterns and climate.
5. **Think Critically** What would happen to temperatures on Earth if the Sun's heat were not distributed throughout the atmosphere?

Applying Math

6. **Solve One-Step Equations** Earth is about 150 million km from the Sun. The radiation coming from the Sun travels at 300,000 km/s. How long does it take for radiation from the Sun to reach Earth?

Science Online booki.msscience.com/self_check_quiz

Air Movement

Forming Wind

Earth is mostly rock or land, with three-fourths of its surface covered by a relatively thin layer of water, the oceans. These two areas strongly influence global wind systems. Uneven heating of Earth's surface by the Sun causes some areas to be warmer than others. Recall that warmer air expands, becoming lower in density than the colder air. This causes air pressure to be generally lower where air is heated. Wind is the movement of air from an area of higher pressure to an area of lower pressure.

Heated Air Areas of Earth receive different amounts of radiation from the Sun because Earth is curved. **Figure 15** illustrates why the equator receives more radiation than areas to the north or south. The heated air at the equator is less dense, so it is displaced by denser, colder air, creating convection currents.

This cold, denser air comes from the poles, which receive less radiation from the Sun, making air at the poles much cooler. The resulting dense, high-pressure air sinks and moves along Earth's surface. However, dense air sinking as less-dense air rises does not explain everything about wind.

as you read

What You'll Learn

- **Explain** why different latitudes on Earth receive different amounts of solar energy.
- **Describe** the Coriolis effect.
- **Explain** how land and water surfaces affect the overlying air.

Why It's Important

Wind systems determine major weather patterns on Earth.

Review Vocabulary
density: mass per unit volume

New Vocabulary
- Coriolis effect
- jet stream
- sea breeze
- land breeze

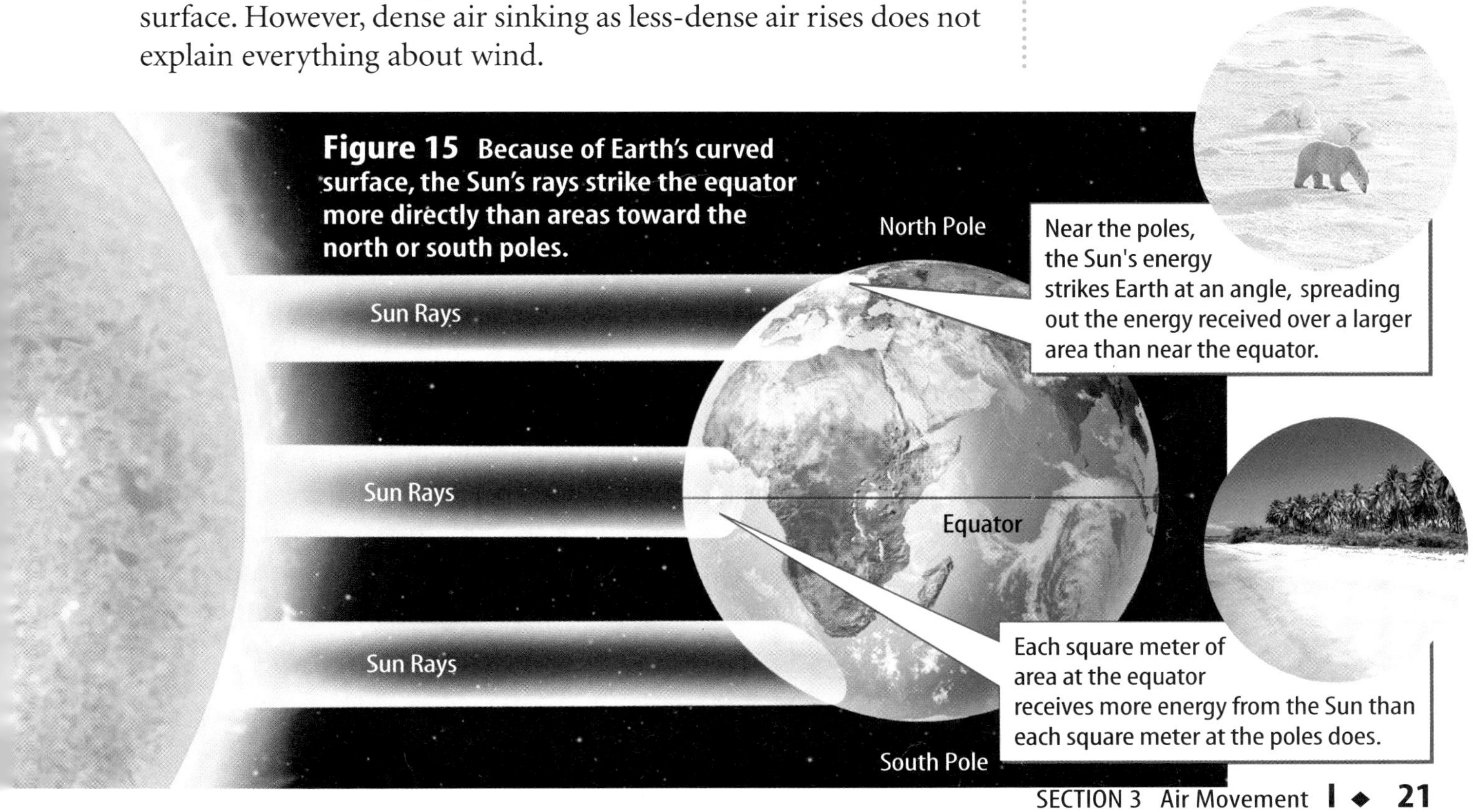

Figure 15 Because of Earth's curved surface, the Sun's rays strike the equator more directly than areas toward the north or south poles.

North Pole

Sun Rays

Sun Rays

Sun Rays

Equator

South Pole

Near the poles, the Sun's energy strikes Earth at an angle, spreading out the energy received over a larger area than near the equator.

Each square meter of area at the equator receives more energy from the Sun than each square meter at the poles does.

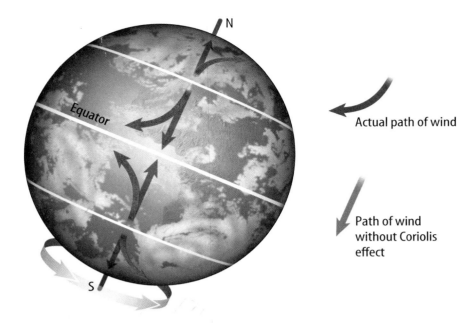

Figure 16 The Coriolis effect causes moving air to turn to the right in the northern hemisphere and to the left in the southern hemisphere.
Explain *What causes this to happen?*

Actual path of wind

Path of wind without Coriolis effect

The Coriolis Effect What would happen if you threw a ball to someone sitting directly across from you on a moving merry-go-round? Would the ball go to your friend? By the time the ball got to the opposite side, your friend would have moved and the ball would appear to have curved.

Like the merry-go-round, the rotation of Earth causes moving air and water to appear to turn to the right north of the equator and to the left south of the equator. This is called the **Coriolis** (kohr ee OH lus) **effect.** It is illustrated in **Figure 16.** The flow of air caused by differences in the amount of solar radiation received on Earth's surface and by the Coriolis effect creates distinct wind patterns on Earth's surface. These wind systems not only influence the weather, they also determine when and where ships and planes travel most efficiently.

Global Winds

How did Christopher Columbus get from Spain to the Americas? The *Nina,* the *Pinta,* and the *Santa Maria* had no source of power other than the wind in their sails. Early sailors discovered that the wind patterns on Earth helped them navigate the oceans. These wind systems are shown in **Figure 17.**

Sometimes sailors found little or no wind to move their sailing ships near the equator. It also rained nearly every afternoon. This windless, rainy zone near the equator is called the doldrums. Look again at **Figure 17.** Near the equator, the Sun heats the air and causes it to rise, creating low pressure and little wind. The rising air then cools, causing rain.

Topic: Global Winds
Visit booki.msscience.com for Web links to information about global winds.

Activity Make a model of Earth showing the locations of global wind patterns.

 **Reading Check** *What are the doldrums?*

Figure 17

The Sun's uneven heating of Earth's surface forms giant loops, or cells, of moving air. The Coriolis effect deflects the surface winds to the west or east, setting up belts of prevailing winds that distribute heat and moisture around the globe.

A WESTERLIES Near 30° north and south latitude, Earth's rotation deflects air from west to east as air moves toward the polar regions. In the United States, the westerlies move weather systems, such as this one along the Oklahoma-Texas border, from west to east.

B DOLDRUMS Along the equator, heating causes air to expand, creating a zone of low pressure. Cloudy, rainy weather, as shown here, develops almost every afternoon.

60° N — Polar easterlies
Westerlies
30° N —
Trade winds
0° — Equatorial doldrums
Trade winds
30° S —
Westerlies
60° S — Polar easterlies

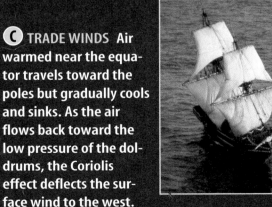

C TRADE WINDS Air warmed near the equator travels toward the poles but gradually cools and sinks. As the air flows back toward the low pressure of the doldrums, the Coriolis effect deflects the surface wind to the west. Early sailors, in ships like the one above, relied on these winds to navigate global trade routes.

D POLAR EASTERLIES In the polar regions, cold, dense air sinks and moves away from the poles. Earth's rotation deflects this wind from east to west.

Surface Winds Air descending to Earth's surface near 30° north and south latitude creates steady winds that blow in tropical regions. These are called trade winds because early sailors used their dependability to establish trade routes.

Between 30° and 60° latitude, winds called the prevailing westerlies blow in the opposite direction from the trade winds. Prevailing westerlies are responsible for much of the movement of weather across North America.

Polar easterlies are found near the poles. Near the north pole, easterlies blow from northeast to southwest. Near the south pole, polar easterlies blow from the southeast to the northwest.

Winds in the Upper Troposphere Narrow belts of strong winds, called **jet streams,** blow near the top of the troposphere. The polar jet stream forms at the boundary of cold, dry polar air to the north and warmer, more moist air to the south, as shown in **Figure 18.** The jet stream moves faster in the winter because the difference between cold air and warm air is greater. The jet stream helps move storms across the country.

Jet pilots take advantage of the jet streams. When flying eastward, planes save time and fuel. Going west, planes fly at different altitudes to avoid the jet streams.

Local Wind Systems

Global wind systems determine the major weather patterns for the entire planet. Smaller wind systems affect local weather. If you live near a large body of water, you're familiar with two such wind systems—sea breezes and land breezes.

Figure 18 The polar jet stream affecting North America forms along a boundary where colder air lies to the north and warmer air lies to the south. It is a swiftly flowing current of air that moves in a wavy west-to-east direction and is usually found between 10 km and 15 km above Earth's surface.

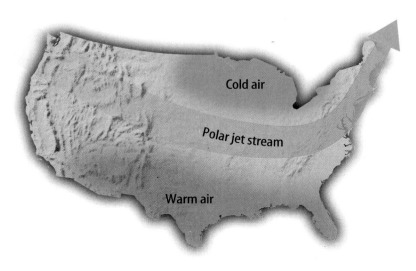

Cold air

Polar jet stream

Warm air

Flying from Boston to Seattle may take 30 min longer than flying from Seattle to Boston.
Think Critically *Why would it take longer to fly from east to west than it would from west to east?*

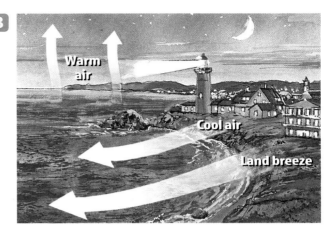

Sea and Land Breezes Convection currents over areas where the land meets the sea can cause wind. A **sea breeze,** shown in **Figure 19,** is created during the day because solar radiation warms the land more than the water. Air over the land is heated by conduction. This heated air is less dense and has lower pressure. Cooler, denser air over the water has higher pressure and flows toward the warmer, less dense air. A convection current results, and wind blows from the sea toward the land. The reverse occurs at night, when land cools much more rapidly than ocean water. Air over the land becomes cooler than air over the ocean. Cooler, denser air above the land moves over the water, as the warm air over the water rises. Movement of air toward the water from the land is called a **land breeze.**

Figure 19 These daily winds occur because land heats up and cools off faster than water does. **A** During the day, cool air from the water moves over the land, creating a sea breeze. **B** At night, cool air over the land moves toward the warmer air over the water, creating a land breeze.

✔ **Reading Check** *How does a sea breeze form?*

section 3 review

Summary

Forming Wind

- Warm air is less dense than cool air.
- Differences in density and pressure cause air movement and wind.
- The Coriolis effect causes moving air to appear to turn right north of the equator and left south of the equator.

Wind Systems

- Wind patterns are affected by latitude.
- High-altitude belts of wind, called jet streams, can be found near the top of the troposphere.
- Sea breezes blow from large bodies of water toward land, while land breezes blow from land toward water.

Self Check

1. **Conclude** why some parts of Earth's surface, such as the equator, receive more of the Sun's heat than other regions.
2. **Explain** how the Coriolis effect influences winds.
3. **Analyze** why little wind and much afternoon rain occur in the doldrums.
4. **Infer** which wind system helped early sailors navigate Earth's oceans.
5. **Think Critically** How does the jet stream help move storms across North America?

Applying Skills

6. **Compare and contrast** sea breezes and land breezes.

The Heat Is On

▶ Real-World Question

Sometimes, a plunge in a pool or lake on a hot summer day feels cool and refreshing. Why does the beach sand get so hot when the water remains cool? A few hours later, the water feels warmer than the land does. How do soil and water compare in their abilities to absorb and emit heat?

▶ Form a Hypothesis

Form a hypothesis about how soil and water compare in their abilities to absorb and release heat. Write another hypothesis about how air temperatures above soil and above water differ during the day and night.

Goals

■ **Design** an experiment to compare heat absorption and release for soil and water.
■ **Observe** how heat release affects the air above soil and above water.

Possible Materials

ring stand
soil
metric ruler
water
masking tape
clear-plastic boxes (2)
overhead light
    with reflector
thermometers (4)
colored pencils (4)

Safety Precautions

WARNING: *Be careful when handling the hot overhead light. Do not let the light or its cord make contact with water.*

● Test Your Hypothesis

Make a Plan

1. As a group, agree upon and write your hypothesis.

2. **List** the steps that you need to take to test your hypothesis. Include in your plan a description of how you will use your equipment to compare heat absorption and release for water and soil.

3. **Design** a data table in your Science Journal for both parts of your experiment—when the light is on and energy can be absorbed and when the light is off and energy is released to the environment.

Follow Your Plan

1. Make sure your teacher approves your plan and your data table before you start.

2. Carry out the experiment as planned.

3. During the experiment, record your observations and complete the data table in your Science Journal.

4. Include the temperatures of the soil and the water in your measurements. Also compare heat release for water and soil. Include the temperatures of the air immediately above both of the substances. Allow 15 min for each test.

● Analyze Your Data

1. Use your colored pencils and the information in your data tables to make line graphs. Show the rate of temperature increase for soil and water. Graph the rate of temperature decrease for soil and water after you turn the light off.

2. **Analyze** your graphs. When the light was on, which heated up faster—the soil or the water?

3. **Compare** how fast the air temperature over the water changed with how fast the temperature over the land changed after the light was turned off.

● Conclude and Apply

1. Were your hypotheses supported or not? Explain.

2. **Infer** from your graphs which cooled faster—the water or the soil.

3. **Compare** the temperatures of the air above the water and above the soil 15 minutes after the light was turned off. How do water and soil compare in their abilities to absorb and release heat?

Communicating Your Data

Make a poster showing the steps you followed for your experiment. Include graphs of your data. Display your poster in the classroom.

Song of the Sky Loom[1]

Brian Swann, ed.

This Native American prayer probably comes from the Tewa-speaking Pueblo village of San Juan, New Mexico. The poem is actually a chanted prayer used in ceremonial rituals.

Mother Earth   Father Sky

we are your children
With tired backs we bring you gifts you love
Then weave for us a garment of brightness
its warp[2] the white light of morning,
weft[3] the red light of evening,
fringes the falling rain,
its border the standing rainbow.
Thus weave for us a garment of brightness
So we may walk fittingly where birds sing,
So we may walk fittingly where grass is green.

Mother Earth   Father Sky

1 a machine or device from which cloth is produced

2 threads that run lengthwise in a piece of cloth

3 horizontal threads interlaced through the warp in a piece of cloth

Understanding Literature

Metaphor A metaphor is a figure of speech that compares seemingly unlike things. Unlike a simile, a metaphor does not use the connecting words *like* or *as*. Why does the song use the image of a garment to describe Earth's atmosphere?

Respond to Reading

1. What metaphor does the song use to describe Earth's atmosphere?
2. Why do the words *Mother Earth* and *Father Sky* appear on either side and above and below the rest of the words?
3. **Linking Science and Writing** Write a four-line poem that uses a metaphor to describe rain.

 In this chapter, you learned about the composition of Earth's atmosphere. The atmosphere maintains the proper balance between the amount of heat absorbed from the Sun and the amount of heat that escapes back into space. The water cycle explains how water evaporates from Earth's surface back into the atmosphere. Using metaphor instead of scientific facts, the Tewa song conveys to the reader how the relationship between Earth and its atmosphere is important to all living things.

Reviewing Main Ideas

Section 1 — Earth's Atmosphere

1. Earth's atmosphere is made up mostly of gases, with some suspended solids and liquids. The unique atmosphere allows life on Earth to exist.

2. The atmosphere is divided into five layers with different characteristics.

3. The ozone layer protects Earth from too much ultraviolet radiation, which can be harmful.

Section 2 — Energy Transfer in the Atmosphere

1. Earth receives its energy from the Sun. Some of this energy is reflected back into space, and some is absorbed.

2. Heat is distributed in Earth's atmosphere by radiation, conduction, and convection.

3. Energy from the Sun powers the water cycle between the atmosphere and Earth's surface.

4. Unlike the atmosphere on Mars or Venus, Earth's unique atmosphere maintains a balance between energy received and energy lost that keeps temperatures mild. This delicate balance allows life on Earth to exist.

Section 3 — Air Movement

1. Because Earth's surface is curved, not all areas receive the same amount of solar radiation. This uneven heating causes temperature differences at Earth's surface.

2. Convection currents modified by the Coriolis effect produce Earth's global winds.

3. The polar jet stream is a strong current of wind found in the upper troposphere. It forms at the boundary between cold, polar air and warm, tropical air.

4. Land breezes and sea breezes occur near the ocean.

Visualizing Main Ideas

Copy and complete the following cycle map on the water cycle.

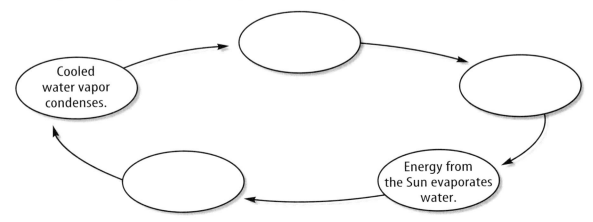

Cooled water vapor condenses.

Energy from the Sun evaporates water.

Using Vocabulary

atmosphere p. 8	jet stream p. 24
chlorofluorocarbon p. 14	land breeze p. 25
condensation p. 19	ozone layer p. 14
conduction p. 18	radiation p. 18
convection p. 18	sea breeze p. 25
Coriolis effect p. 22	troposphere p. 10
hydrosphere p. 19	ultraviolet radiation p. 14
ionosphere p. 11	

Fill in the blanks below with the correct vocabulary word or words.

1. Chlorofluorocarbons are dangerous because they destroy the _____.

2. Narrow belts of strong winds called _____ blow near the top of the troposphere.

3. The thin layer of air that surrounds Earth is called the _____.

4. Heat energy transferred in the form of waves is called _____.

5. The ozone layer helps protect us from _____.

Checking Concepts

Choose the word or phrase that best answers the question.

6. Nitrogen makes up what percentage of the atmosphere?
   A) 21%   C) 78%
   B) 1%   D) 90%

7. What causes a brown haze near cities?
   A) conduction
   B) mud
   C) car exhaust
   D) wind

8. Which is the uppermost layer of the atmosphere?
   A) troposphere   C) exosphere
   B) stratosphere   D) thermosphere

9. What layer of the atmosphere has the most water?
   A) troposphere   C) mesosphere
   B) stratosphere   D) exosphere

10. What protects living things from too much ultraviolet radiation?
    A) the ozone layer   C) nitrogen
    B) oxygen   D) argon

11. Where is air pressure least?
    A) troposphere   C) exosphere
    B) stratosphere   D) thermosphere

12. How is energy transferred when objects are in contact?
    A) trade winds   C) radiation
    B) convection   D) conduction

13. Which surface winds are responsible for most of the weather movement across the United States?
    A) polar easterlies
    B) sea breeze
    C) prevailing westerlies
    D) trade winds

14. What type of wind is a movement of air toward water?
    A) sea breeze
    B) polar easterlies
    C) land breeze
    D) trade winds

15. What are narrow belts of strong winds near the top of the troposphere called?
    A) doldrums
    B) jet streams
    C) polar easterlies
    D) trade winds

Thinking Critically

16. **Explain** why there are few or no clouds in the stratosphere.

17. **Describe** It is thought that life could not have existed on land until the ozone layer formed about 2 billion years ago. Why does life on land require an ozone layer?

18. **Diagram** Why do sea breezes occur during the day but not at night?

19. **Describe** what happens when water vapor rises and cools.

20. **Explain** why air pressure decreases with an increase in altitude.

21. **Concept Map** Copy and complete the cycle concept map below using the following phrases to explain how air moves to form a convection current: *Cool air moves toward warm air, warm air is lifted and cools, and cool air sinks.*

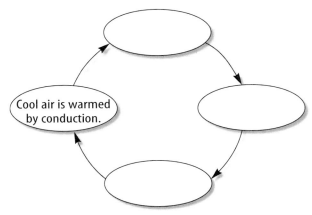

Cool air is warmed by conduction.

22. **Form Hypotheses** Carbon dioxide in the atmosphere prevents some radiation from Earth's surface from escaping to space. Hypothesize how the temperature on Earth might change if more carbon dioxide were released from burning fossil fuels.

23. **Identify and Manipulate Variables and Controls** Design an experiment to find out how plants are affected by differing amounts of ultraviolet radiation. In the design, use filtering film made for car windows. What is the variable you are testing? What are your constants? Your controls?

24. **Recognize Cause and Effect** Why is the inside of a car hotter than the outdoor temperature on a sunny summer day?

Performance Activities

25. **Make a Poster** Find newspaper and magazine photos that illustrate how the water cycle affects weather patterns and climate around the world.

26. **Experiment** Design and conduct an experiment to find out how different surfaces such as asphalt, soil, sand, and grass absorb and reflect solar energy. Share the results with your class.

Applying Math

Use the graph below to answer questions 27–28.

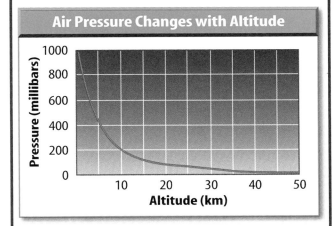

27. **Altitude and Air Pressure** What is the altitude at which air pressure is about 1,000 millibars? What is it at 200 millibars?

28. **Mt. Everest** Assume the altitude on Mt. Everest is about 10 km high. How many times greater is air pressure at sea level than on top of Mt. Everest?

Part 1 | Multiple Choice

Record your answers on the answer sheet provided by your teacher or on a sheet of paper.

Use the illustration below to answer questions 1–3.

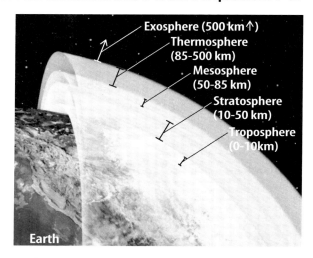

Exosphere (500 km↑)
Thermosphere (85-500 km)
Mesosphere (50-85 km)
Stratosphere (10-50 km)
Troposphere (0-10km)
Earth

1. Which layer of the atmosphere contains the ozone layer?
   A. exosphere
   B. mesosphere
   C. stratosphere
   D. troposphere

2. Which atmospheric layer contains weather?
   A. mesosphere
   B. stratosphere
   C. thermosphere
   D. troposphere

3. Which atmospheric layer contains electrically charged particles?
   A. stratosphere
   B. ionosphere
   C. exosphere
   D. troposphere

4. What process changes water vapor to a liquid?
   A. condensation
   B. evaporation
   C. infiltration
   D. precipitation

5. Which process transfers heat by contact?
   A. conduction
   B. convection
   C. evaporation
   D. radiation

6. Which global wind affects weather in the U.S.?
   A. doldrums   C. trade winds
   B. easterlies   D. westerlies

Use the illustration below to answer question 7.

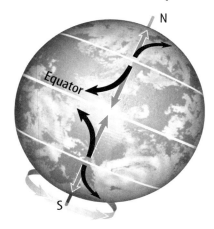

N
Equator
S

7. Which deflects winds to the west or east?
   A. convection
   B. Coriolis effect
   C. jet stream
   D. radiation

8. Which forms during the day because water heats slower than land?
   A. easterlies   C. land breeze
   B. westerlies   D. sea breeze

9. Which is the most abundant gas in Earth's atmosphere?
   A. carbon dioxide
   B. nitrogen
   C. oxygen
   D. water vapor

Part 2 | Short Response/Grid In

Record your answers on the answer sheet provided by your teacher or on a sheet of paper.

10. Why does pressure drop as you travel upward from Earth's surface?

11. Why does the equator receive more radiation than areas to the north or south?

12. Why does a land breeze form at night?

13. Why does the jet stream move faster in the winter?

14. Why is one global wind pattern known as the trade winds?

Use the illustration below to answer questions 15–17.

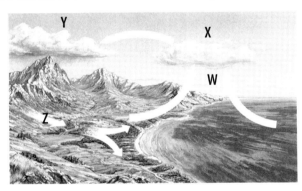

15. What process is illustrated?

16. Explain how this cycle affects weather patterns and climate.

17. What happens to water that falls as precipitation and does not runoff and flow into streams?

18. How do solid particles become part of Earth's atmosphere?

19. Why can flying from Seattle to Boston take less time than flying from Boston to Seattle in the same aircraft?

Part 3 | Open Ended

Record your answers on a sheet of paper.

20. Explain how ozone is destroyed by chlorofluorocarbons.

21. Explain how Earth can heat the air by conduction.

22. Explain how humans influence the composition of Earth's atmosphere.

23. Draw three diagrams to demonstrate radiation, convection, and conduction.

24. Explain why the doldrums form over the equator.

Part 2 | Constructed Response

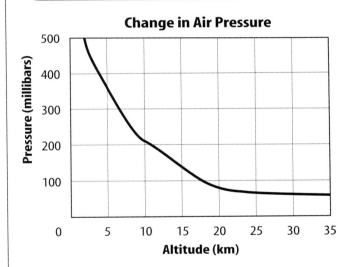

Change in Air Pressure

25. As you increase in altitude what happens to the air pressure? How might this affect people who move to the mountains?

Test-Taking Tip

Trends in Graphs When analyzing data in a table or graph, look for a trend. Questions about the pattern may use words like *increase, decrease, hypothesis,* or *summary.*

Question 25 The word "increase" indicates that you should look for the trend in air pressure as altitude increases.

Weather

The BIG Idea

Weather refers to the state of the atmosphere at a specific time and place.

SECTION 1
What is weather?
Main Idea Weather describes conditions such as air pressure, wind, temperature, and the amount of moisture in the air.

SECTION 2
Weather Patterns
Main Idea Because of the movement of air and moisture in the atmosphere, weather constantly changes.

SECTION 3
Weather Forecasts
Main Idea Meteorologists gather information about current weather to make predictions about future weather patterns.

To play or not to play?
Will this approaching storm be over before the game begins? New weather technology can provide information that allows us to make plans based on predicted weather conditions, such as whether or not to delay the start of a baseball game.

Science Journal Write three questions you would ask a meteorologist about weather.

Start-Up Activities

What causes rain?

How can it rain one day and be sunny the next? Powered by heat from the Sun, the air that surrounds you stirs and swirls. This constant mixing produces storms, calm weather, and everything in between. What causes rain and where does the water come from? Do the lab below to find out. **WARNING:** *Boiling water and steam can cause burns.*

1. Bring a pan of water to a boil on a hot plate.
2. Carefully hold another pan containing ice cubes about 20 cm above the boiling water. Be sure to keep your hands and face away from the steam.
3. Keep the pan with the ice cubes in place until you see drops of water dripping from the bottom.
4. **Think Critically** In your Science Journal, describe how the droplets formed. Infer where the water on the bottom of the pan came from.

Weather When information is grouped into clear categories, it is easier to make sense of what you are learning. Make the following Foldable to help you organize your thoughts about weather.

STEP 1 Collect 2 sheets of paper and layer them about 1.25 cm apart vertically. Keep the edges level.

STEP 2 Fold up the bottom edges of the paper to form 4 equal tabs.

STEP 3 Fold the papers and crease well to hold the tabs in place. Staple along the fold.

STEP 4 Label the tabs *Weather, What is weather?, Weather Patterns,* and *Forecasting Weather* as shown.

Summarize As you read the chapter, summarize what you learn under the appropriate tabs.

 Preview this chapter's content and activities at booki.msscience.com

Get Ready to Read

New Vocabulary

① Learn It! What should you do if you find a word you don't know or understand? Here are some suggested strategies:

1. Use context clues (from the sentence or the paragraph) to help you define it.
2. Look for prefixes, suffixes, or root words that you already know.
3. Write it down and ask for help with the meaning.
4. Guess at its meaning.
5. Look it up in the glossary or a dictionary.

② Practice It! Look at the phrase *air mass* in the following passage. See how context clues can help you understand its meaning.

Context Clue
Air masses can vary in the amount of moisture content and can vary in temperature.

Context Clue
Air masses can be very large.

Context Clue
Air masses can move.

. . . an air mass that develops over land is dry compared with one that develops over water. An air mass that develops in the tropics is warmer than one that develops over northern regions. An air mass can cover thousands of square kilometers. When you observe a change in the weather from one day to the next, it is due to the movement of air masses.

—*from page 44*

③ Apply It! Make a vocabulary bookmark with a strip of paper. As you read, keep track of words you do not know or want to learn more about.

Target Your Reading

Reading Tip Read a paragraph containing a vocabulary word from beginning to end. Then, go back to determine the meaning of the word.

Use this to focus on the main ideas as you read the chapter.

1. **Before you read** the chapter, respond to the statements below on your worksheet or on a numbered sheet of paper.
   - Write an **A** if you **agree** with the statement.
   - Write a **D** if you **disagree** with the statement.

2. **After you read** the chapter, look back to this page to see if you've changed your mind about any of the statements.
   - If any of your answers changed, explain why.
   - Change any false statements into true statements.
   - Use your revised statements as a study guide.

Science nline

Print out a worksheet of this page at booki.msscience.com

Before You Read A or D		Statement	After You Read A or D
	1	Heat from the Sun is absorbed by Earth's surface, which then heats the air above it.	
	2	Clouds are made of frozen air.	
	3	Dew point is a specific temperature and is not related to the amount of moisture in the air.	
	4	Clouds are classified by shape and height.	
	5	Cold fronts occur only in northern climates, and warm fronts occur only in southern climates.	
	6	Lightning can occur within a cloud, between clouds, or between a cloud and the ground.	
	7	Thunder is caused by clouds crashing into each other.	
	8	Hurricanes gain strength from heat and moisture of warm ocean water.	
	9	A meteorologist studies weather.	
	10	An isobar is a cold, icy, iron bar.	

What is weather?

as you read

What You'll Learn

- **Explain** how solar heating and water vapor in the atmosphere affect weather.
- **Discuss** how clouds form and how they are classified.
- **Describe** how rain, hail, sleet, and snow develop.

Why It's Important

Weather changes affect your daily activities.

Review Vocabulary

factor: something that influences a result

New Vocabulary

- weather
- humidity
- relative humidity
- dew point
- fog
- precipitation

Weather Factors

It might seem like small talk to you, but for farmers, truck drivers, pilots, and construction workers, the weather can have a huge impact on their livelihoods. Even professional athletes, especially golfers, follow weather patterns closely. You can describe what happens in different kinds of weather, but can you explain how it happens?

Weather refers to the state of the atmosphere at a specific time and place. Weather describes conditions such as air pressure, wind, temperature, and the amount of moisture in the air.

The Sun provides almost all of Earth's energy. Energy from the Sun evaporates water into the atmosphere where it forms clouds. Eventually, the water falls back to Earth as rain or snow. However, the Sun does more than evaporate water. It is also a source of heat energy. Heat from the Sun is absorbed by Earth's surface, which then heats the air above it. Differences in Earth's surface lead to uneven heating of Earth's atmosphere. Heat is eventually redistributed by air and water currents. Weather, as shown in **Figure 1,** is the result of heat and Earth's air and water.

Figure 1 The Sun provides the energy that drives Earth's weather.
Identify *storms in this image.*

Molecules in air

Wind

Temperature Pressure

Molecules in air

Temperature Pressure

When air is heated, it expands and becomes less dense. This creates lower pressure.

Molecules making up air are closer together in cooler temperatures, creating high pressure. Wind blows from higher pressure toward lower pressure.

Figure 2 The temperature of air can affect air pressure. Wind is air moving from high pressure to low pressure.

Infer *In the above picture, which way would the wind move at night if the land cooled?*

Air Temperature During the summer when the Sun is hot and the air is still, a swim can be refreshing. But would a swim seem refreshing on a cold, winter day? The temperature of air influences your daily activities.

Air is made up of molecules that are always moving randomly, even when there's no wind. Temperature is a measure of the average amount of motion of molecules. When the temperature is high, molecules in air move rapidly and it feels warm. When the temperature is low, molecules in air move less rapidly, and it feels cold.

Wind Why can you fly a kite on some days but not others? Kites fly because air is moving. Air moving in a specific direction is called wind. As the Sun warms Earth's surface, air near the surface is heated by conduction. The air expands, becomes less dense, and rises. Warm, rising air has low atmospheric pressure. Cool, dense air tends to sink, bringing about high atmospheric pressure. Wind results because air moves from areas of high pressure to areas of low pressure. You may have experienced this if you've ever spent time along a beach, as in **Figure 2.**

Many instruments are used to measure wind direction and speed. Wind direction can be measured using a wind vane. A wind vane has an arrow that points in the direction from which the wind is blowing. A wind sock has one open end that catches the wind, causing the sock to point in the direction toward which the wind is blowing. Wind speed can be measured using an anemometer (a nuh MAH muh tur). Anemometers have rotating cups that spin faster when the wind is strong.

INTEGRATE
Life Science

Body Temperature Birds and mammals maintain a fairly constant internal temperature, even when the temperature outside their bodies changes. On the other hand, the internal temperature of fish and reptiles changes when the temperature around them changes. Infer from this which group is more likely to survive a quick change in the weather.

Figure 3 Warmer air can have more water vapor than cooler air can because water vapor doesn't easily condense in warm air.

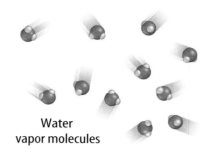

Water vapor molecules

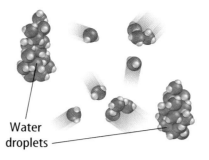

Water droplets

Water vapor molecules in warm air move rapidly. The molecules can't easily come together and condense.

As air cools, water molecules in air move closer together. Some of them collide, allowing condensation to take place.

Mini LAB

Determining Dew Point

Procedure 📧

1. Partially fill a **metal can** with **room-temperature water**. Dry the outer surface of the can.
2. Place a **stirring rod** in the water.
3. Slowly stir the water and add small amounts of **ice**.
4. Make a data table in your **Science Journal**. With a **thermometer,** note the exact water temperature at which a thin film of moisture first begins to form on the outside of the metal can.
5. Repeat steps 1 through 4 two more times.
6. The average of the three temperatures at which the moisture begins to appear is the dew point temperature of the air surrounding the metal container.

Analysis

1. What determines the dew point temperature?
2. Will the dew point change with increasing temperature if the amount of moisture in the air doesn't change? Explain.

Humidity Heat evaporates water into the atmosphere. Where does the water go? Water vapor molecules fit into spaces among the molecules that make up air. The amount of water vapor present in the air is called **humidity.**

Air doesn't always contain the same amount of water vapor. As you can see in **Figure 3,** more water vapor can be present when the air is warm than when it is cool. At warmer temperatures, the molecules of water vapor in air move quickly and don't easily come together. At cooler temperatures, molecules in air move more slowly. The slower movement allows water vapor molecules to stick together and form droplets of liquid water. The formation of liquid water from water vapor is called condensation. When enough water vapor is present in air for condensation to take place, the air is saturated.

✓ Reading Check *Why can more water vapor be present in warm air than in cold air?*

Relative Humidity On a hot, sticky afternoon, the weather forecaster reports that the humidity is 50 percent. How can the humidity be low when it feels so humid? Weather forecasters report the amount of moisture in the air as relative humidity. **Relative humidity** is a measure of the amount of water vapor present in the air compared to the amount needed for saturation at a specific temperature.

If you hear a weather forecaster say that the relative humidity is 50 percent, it means that the air contains 50 percent of the water needed for the air to be saturated.

As shown in **Figure 4,** air at 25°C is saturated when it contains 22 g of water vapor per cubic meter of air. The relative humidity is 100 percent. If air at 25°C contains 11 g of water vapor per cubic meter, the relative humidity is 50 percent.

Dew Point

When the temperature drops, less water vapor can be present in air. The water vapor in air will condense to a liquid or form ice crystals. The temperature at which air is saturated and condensation forms is the dew point. The **dew point** changes with the amount of water vapor in the air.

You've probably seen water droplets form on the outside of a glass of cold milk. The cold glass cooled the air next to it to its dew point. The water vapor in the surrounding air condensed and formed water droplets on the glass. In a similar way, when air near the ground cools to its dew point, water vapor condenses and forms dew. Frost may form when temperatures are near 0°C.

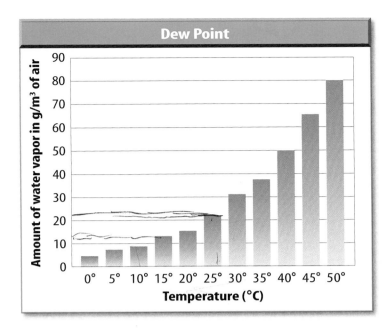

Figure 4 This graph shows that as the temperature of air increases, more water vapor can be present in the air.

Applying Math Calculate Percent

DEW POINT One summer day, the relative humidity is 80 percent and the temperature is 35°C. Use Figure 4 to find the dew point reached if the temperature falls to 25°C?

Solution

1 *This is what you know:*

Air Temperature (°C)	Amount of Water Vapor Needed for Saturation (g/m³)
35	37
25	22
15	14

2 *This is what you need to find out:* x = amount of water vapor in 35°C air at 80 percent relative humidity. Is $x > 22$ g/m³ or is $x < 22$ g/m³?

3 *This is how you solve the problem:*

$x = .80 \, (37 \text{ g/m}^3)$
$x = 29.6$ g/m³ of water vapor
29.6 g/m³ > 22 g/m³, so the dew point is reached and dew will form.

Practice Problems

1. If the relative humidity is 50 percent and the air temperature is 35°C, will the dew point be reached if the temperature falls to 20°C?

2. If the air temperature is 25°C and the relative humidity is 30 percent, will the dew point be reached if the temperature drops to 15°C?

 For more practice, visit booki.msscience.com/ math_practice

Forming Clouds

Why are there clouds in the sky? Clouds form as warm air is forced upward, expands, and cools. **Figure 5** shows several ways that warm, moist air forms clouds. As the air cools, the amount of water vapor needed for saturation decreases and the relative humidity increases. When the relative humidity reaches 100 percent, the air is saturated. Water vapor soon begins to condense in tiny droplets around small particles such as dust and salt. These droplets of water are so small that they remain suspended in the air. Billions of these droplets form a cloud.

Classifying Clouds

Clouds are classified mainly by shape and height. Some clouds extend high into the sky, and others are low and flat. Some dense clouds bring rain or snow, while thin, wispy clouds appear on mostly sunny days. The shape and height of clouds vary with temperature, pressure, and the amount of water vapor in the atmosphere.

Figure 5 Clouds form when moist air is lifted and cools. This occurs where air is heated, at mountain ranges, and where cold air meets warm air.

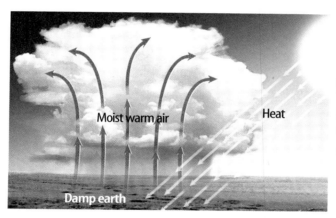

Moist warm air

Heat

Damp earth

Rays from the Sun heat the ground and the air next to it. The warm air rises and cools. If the air is moist, some water vapor condenses and forms clouds.

Moist warm air

As moist air moves over mountains, it is lifted and cools. Clouds formed in this way can cover mountains for long periods of time.

When cool air meets warm, moist air, the warm air is lifted and cools.
Explain *what happens to the water vapor when the dew point is reached.*

Shape The three main cloud types are stratus, cumulus, and cirrus. Stratus clouds form layers, or smooth, even sheets in the sky. Stratus clouds usually form at low altitudes and may be associated with fair weather or rain or snow. When air is cooled to its dew point near the ground, it forms a stratus cloud called **fog,** as shown in **Figure 6.**

Cumulus (KYEW myuh lus) clouds are masses of puffy, white clouds, often with flat bases. They sometimes tower to great heights and can be associated with fair weather or thunderstorms.

Cirrus (SIHR us) clouds appear fibrous or curly. They are high, thin, white, feathery clouds made of ice crystals. Cirrus clouds are associated with fair weather, but they can indicate approaching storms.

Height Some prefixes of cloud names describe the height of the cloud base. The prefix *cirro-* describes high clouds, *alto-* describes middle-elevation clouds, and *strato-* refers to clouds at low elevations. Some clouds' names combine the altitude prefix with the term *stratus* or *cumulus.*

Cirrostratus clouds are high clouds, like those in **Figure 7.** Usually, cirrostratus clouds indicate fair weather, but they also can signal an approaching storm. Altostratus clouds form at middle levels. If the clouds are not too thick, sunlight can filter through them.

Figure 6 Fog surrounds the Golden Gate Bridge, San Francisco. Fog is a stratus cloud near the ground.
Think Critically *Why do you think fog is found in San Francisco Bay?*

Figure 7 Cirrostratus clouds are made of ice crystals and form high in Earth's atmosphere.

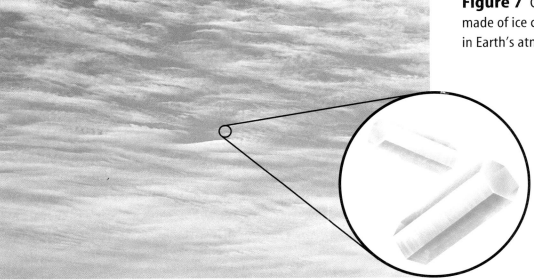
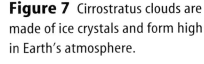

Figure 8 Water vapor in air collects on particles to form water droplets or ice crystals. The type of precipitation that is received on the ground depends on the temperature of the air.

When the air is warm, water vapor forms raindrops that fall as rain.

When the air is cold, water vapor forms snowflakes.

Rain- or Snow-Producing Clouds Clouds associated with rain or snow often have the word nimbus attached to them. The term *nimbus* is Latin for "dark rain cloud" and this is a good description, because the water content of these clouds is so high that little sunlight can pass through them. When a cumulus cloud grows into a thunderstorm, it is called a cumulonimbus (kyew myuh loh NIHM bus) cloud. These clouds can tower to nearly 18 km. Nimbostratus clouds are layered clouds that can bring long, steady rain or snowfall.

Precipitation

Water falling from clouds is called **precipitation.** Precipitation occurs when cloud droplets combine and grow large enough to fall to Earth. The cloud droplets form around small particles, such as salt and dust. These particles are so small that a puff of smoke can contain millions of them.

You might have noticed that raindrops are not all the same size. The size of raindrops depends on several factors. One factor is the strength of updrafts in a cloud. Strong updrafts can keep drops suspended in the air where they can combine with other drops and grow larger. The rate of evaporation as a drop falls to Earth also can affect its size. If the air is dry, the size of raindrops can be reduced or they can completely evaporate before reaching the ground.

Air temperature determines whether water forms rain, snow, sleet, or hail—the four main types of precipitation. **Figure 8** shows these different types of precipitation. Drops of water falling in temperatures above freezing fall as rain. Snow forms when the air temperature is so cold that water vapor changes directly to a solid. Sleet forms when raindrops pass through a layer of freezing air near Earth's surface, forming ice pellets.

 **Reading Check** *What are the four main types of precipitation?*

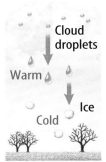

When the air near the ground is cold, sleet, which is made up of many small ice pellets, falls.

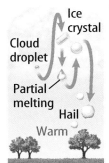

Hailstones are pellets of ice that form inside a cloud.

Hail Hail is precipitation in the form of lumps of ice. Hail forms in cumulonimbus clouds of a thunderstorm when water freezes in layers around a small nucleus of ice. Hailstones grow larger as they're tossed up and down by rising and falling air. Most hailstones are smaller than 2.5 cm but can grow larger than a softball. Of all forms of precipitation, hail produces the most damage immediately, especially if winds blow during a hailstorm. Falling hailstones can break windows and destroy crops.

If you understand the role of water vapor in the atmosphere, you can begin to understand weather. The relative humidity of the air helps determine whether a location will have a dry day or experience some form of precipitation. The temperature of the atmosphere determines the form of precipitation. Studying clouds can add to your ability to forecast weather.

section 1 review

Summary

Weather Factors

- Weather is the state of the atmosphere at a specific time and place.
- Temperature, wind, air pressure, dew point, and humidity describe weather.

Clouds

- Warm, moist air rises, forming clouds.
- The main types of clouds are stratus, cumulus, and cirrus.

Precipitation

- Water falling from clouds is called precipitation.
- Air temperature determines whether water forms rain, snow, sleet, or hail.

Self Check

1. **Explain** When does water vapor in air condense?
2. **Compare and contrast** humidity and relative humidity.
3. **Summarize** how clouds form.
4. **Describe** How does precipitation occur and what determines the type of precipitation that falls to Earth?
5. **Think Critically** Cumulonimbus clouds form when warm, moist air is suddenly lifted. How can the same cumulonimbus cloud produce rain and hail?

Applying Math

6. **Use Graphs** If the air temperature is 30°C and the relative humidity is 60 percent, will the dew point be reached if the temperature drops to 25°C? Use the graph in **Figure 4** to explain your answer.

Weather Patterns

as you read

What **You'll Learn**

- **Describe** how weather is associated with fronts and high- and low-pressure areas.
- **Explain** how tornadoes develop from thunderstorms.
- **Discuss** the dangers of severe weather.

Why **It's Important**

Air masses, pressure systems, and fronts cause weather to change.

🔊 **Review Vocabulary**
barometer: instrument used to measure atmospheric pressure

New Vocabulary
- air mass   • hurricane
- front   • blizzard
- tornado

Weather Changes

When you leave for school in the morning, the weather might be different from what it is when you head home in the afternoon. Because of the movement of air and moisture in the atmosphere, weather constantly changes.

Air Masses   An **air mass** is a large body of air that has properties similar to the part of Earth's surface over which it develops. For example, an air mass that develops over land is dry compared with one that develops over water. An air mass that develops in the tropics is warmer than one that develops over northern regions. An air mass can cover thousands of square kilometers. When you observe a change in the weather from one day to the next, it is due to the movement of air masses. **Figure 9** shows air masses that affect the United States.

Figure 9   Six major air masses affect weather in the United States. Each air mass has the same characteristics of temperature and moisture content as the area over which it formed.

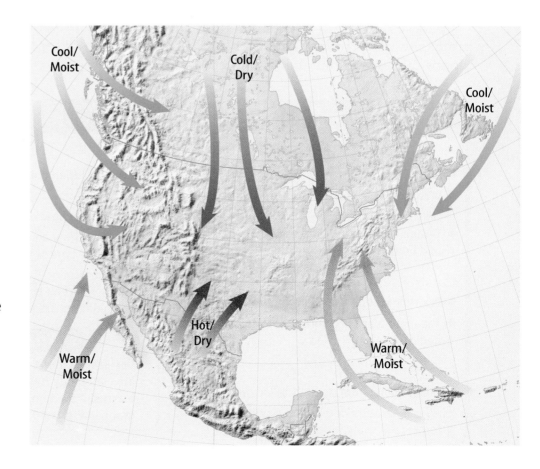

Cool/
Moist

Cold/
Dry

Cool/
Moist

Hot/
Dry

Warm/
Moist

Warm/
Moist

Highs and Lows Atmospheric pressure varies over Earth's surface. Anyone who has watched a weather report on television has heard about high- and low-pressure systems. Recall that winds blow from areas of high pressure to areas of low pressure. As winds blow into a low-pressure area in the northern hemisphere, Earth's rotation causes these winds to swirl in a counterclockwise direction. Large, swirling areas of low pressure are called cyclones and are associated with stormy weather.

✔ Reading Check *How do winds move in a cyclone?*

Winds blow away from a center of high pressure. Earth's rotation causes these winds to spiral clockwise in the northern hemisphere. High-pressure areas are associated with fair weather and are called anticyclones. Air pressure is measured using a barometer, like the one shown in **Figure 10.**

Variation in atmospheric pressure affects the weather. Low pressure systems at Earth's surface are regions of rising air. Clouds form when air is lifted and cools. Areas of low pressure usually have cloudy weather. Sinking motion in high-pressure air masses makes it difficult for air to rise and clouds to form. That's why high pressure usually means good weather.

Figure 10 A barometer measures atmospheric pressure. The red pointer points to the current pressure. Watch how atmospheric pressure changes over time when you line up the white pointer to the one indicating the current pressure each day.

Fronts

A boundary between two air masses of different density, moisture, or temperature is called a **front.** If you've seen a weather map in the newspaper or on the evening news, you've seen fronts represented by various types of curving lines.

Cloudiness, precipitation, and storms sometimes occur at frontal boundaries. Four types of fronts include cold, warm, occluded, and stationary.

Cold and Warm Fronts A cold front, shown on a map as a blue line with triangles ▲▲▲, occurs when colder air advances toward warm air. The cold air wedges under the warm air like a plow. As the warm air is lifted, it cools and water vapor condenses, forming clouds. When the temperature difference between the cold and warm air is large, thunderstorms and even tornadoes may form.

Warm fronts form when lighter, warmer air advances over heavier, colder air. A warm front is drawn on weather maps as a red line with red semicircles .

Science☉nline

Topic: Atmospheric Pressure
Visit booki.msscience.com for Web links to information about the current atmospheric pressure of your town or nearest city.

Activity Look up the pressure of a city west of your town and the pressure of a city to the east. Compare the pressures to local weather conditions. Share your information with the class.

Occluded and Stationary Fronts An occluded front involves three air masses of different temperatures—colder air, cool air, and warm air. An occluded front may form when a cold air mass moves toward cool air with warm air between the two. The colder air forces the warm air upward, closing off the warm air from the surface. Occluded fronts are shown on maps as purple lines with triangles and semicircles ▲●▲.

A stationary front occurs when a boundary between air masses stops advancing. Stationary fronts may remain in the same place for several days, producing light wind and precipitation. A stationary front is drawn on a weather map as an alternating red and blue line. Red semicircles point toward the cold air and blue triangles point toward the warm air ◠▾◠. **Figure 11** summarizes the four types of fronts.

Figure 11 Cold, warm, occluded, and stationary fronts occur at the boundaries of air masses.
Describe *what type of weather occurs at front boundaries.*

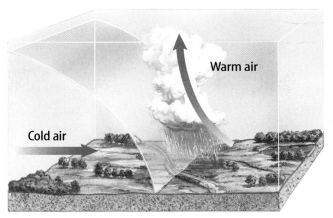

A cold front can advance rapidly. Thunderstorms often form as warm air is suddenly lifted up over the cold air.

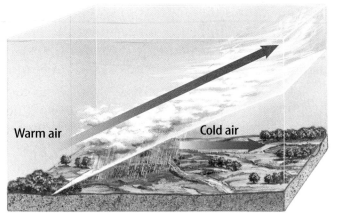

Warm air slides over colder air along a warm front, forming a boundary with a gentle slope. This can lead to hours, if not days, of wet weather.

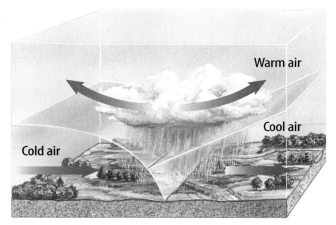

The term *occlusion* means "closure." Colder air forces warm air upward, forming an occluded front that closes off the warm air from the surface.

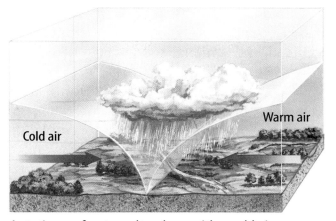

A stationary front results when neither cold air nor warm air advances.

Severe Weather

Despite the weather, you usually can do your daily activities. If it's raining, you still go to school. You can still get there even if it snows a little. However, some weather conditions, such as those caused by thunderstorms, tornadoes, and blizzards, prevent you from going about your normal routine. Severe weather poses danger to people, structures, and animals.

Thunderstorms In a thunderstorm, heavy rain falls, lightning flashes, thunder roars, and hail might fall. What forces cause such extreme weather conditions?

Thunderstorms occur in warm, moist air masses and along fronts. Warm, moist air can be forced upward where it cools and condensation occurs, forming cumulonimbus clouds that can reach heights of 18 km, like the one in **Figure 12.** When rising air cools, water vapor condenses into water droplets or ice crystals. Smaller droplets collide to form larger ones, and the droplets fall through the cloud toward Earth's surface. The falling droplets collide with still more droplets and grow larger. Raindrops cool the air around them. This cool, dense air then sinks and spreads over Earth's surface. Sinking, rain-cooled air and strong updrafts of warmer air cause the strong winds associated with thunderstorms. Hail also may form as ice crystals alternately fall to warmer layers and are lifted into colder layers by the strong updrafts inside cumulonimbus clouds.

Thunderstorm Damage Sometimes thunderstorms can stall over a region, causing rain to fall heavily for a period of time. When streams cannot contain all the water running into them, flash flooding can occur. Flash floods can be dangerous because they occur with little warning.

Strong winds generated by thunderstorms also can cause damage. If a thunderstorm is accompanied by winds traveling faster than 89 km/h, it is classified as a severe thunderstorm. Hail from a thunderstorm can dent cars and the aluminum siding on houses. Although rain from thunderstorms helps crops grow, hail has been known to flatten and destroy entire crops in a matter of minutes.

Figure 12 Tall cumulonimbus clouds may form quickly as warm, moist air rapidly rises.
Identify *some things these clouds are known to produce.*

Figure 13 This time-elapsed photo shows a thunderstorm over Arizona.

Lightning and Thunder

What are lightning and thunder? Inside a storm cloud, warm air is lifted rapidly as cooler air sinks. This movement of air can cause different parts of a cloud to become oppositely charged. When current flows between regions of opposite electrical charge, lightning flashes. Lightning, as shown in **Figure 13,** can occur within a cloud, between clouds, or between a cloud and the ground.

Thunder results from the rapid heating of air around a bolt of lightning. Lightning can reach temperatures of about 30,000°C, which is more than five times the temperature of the surface of the Sun. This extreme heat causes air around the lightning to expand rapidly. Then it cools quickly and contracts. The rapid movement of the molecules forms sound waves heard as thunder.

Tornadoes Some of the most severe thunderstorms produce tornadoes. A **tornado** is a violently rotating column of air in contact with the ground. In severe thunderstorms, wind at different heights blows in different directions and at different speeds. This difference in wind speed and direction, called wind shear, creates a rotating column parallel to the ground. A thunderstorm's updraft can tilt the rotating column upward into the thunderstorm creating a funnel cloud. If the funnel comes into contact with Earth's surface, it is called a tornado.

 **Reading Check** *What causes a tornado to form?*

A tornado's destructive winds can rip apart buildings and uproot trees. High winds can blow through broken windows. When winds blow inside a house, they can lift off the roof and blow out the walls, making it look as though the building exploded. The updraft in the center of a powerful tornado can lift animals, cars, and even houses into the air. Although tornadoes rarely exceed 200 m in diameter and usually last only a few minutes, they often are extremely destructive. In May 1999, multiple thunderstorms produced more than 70 tornadoes in Kansas, Oklahoma, and Texas. This severe tornado outbreak caused 40 deaths, 100 injuries, and more than $1.2 billion in property damage.

Science Online

Topic: Lightning
Visit booki.msscience.com for Web links to research the number of lightning strikes in your state during the last year.

Activity Compare your findings with data from previous years. Communicate to your class what you learn.

Figure 14

Tornadoes are extremely rapid, rotating winds that form at the base of cumulonimbus clouds. Smaller tornadoes may even form inside larger ones. Luckily, most tornadoes remain on the ground for just a few minutes. During that time, however, they can cause considerable—and sometimes strange—damage, such as driving a fork into a tree.

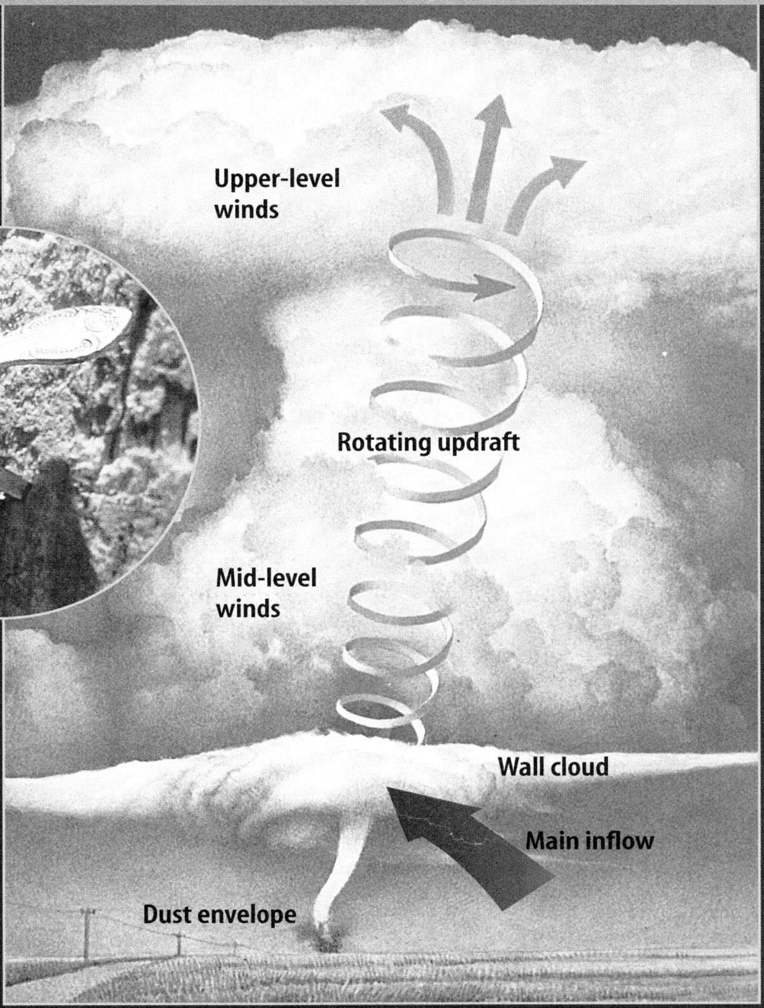

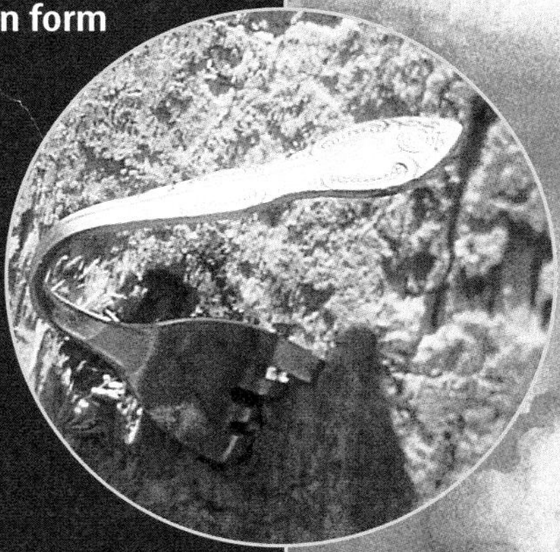

Tornadoes often form from a type of cumulonimbus cloud called a wall cloud. Strong, spiraling updrafts of warm, moist air may form in these clouds. As air spins upward, a low-pressure area forms, and the cloud descends to the ground in a funnel. The tornado sucks up debris as it moves along the ground, forming a dust envelope.

Upper-level winds

Rotating updraft

Mid-level winds

Wall cloud

Main inflow

Dust envelope

The Fujita Scale

	Wind speed (km/h)	Damage
F0	<116	Light: broken branches and chimneys
F1	116–180	Moderate: roofs damaged, mobile homes upturned
F2	181–253	Considerable: roofs torn off homes, large trees uprooted
F3	254–332	Severe: trains overturned, roofs and walls torn off
F4	333–419	Devastating: houses completely destroyed, cars picked up and carried elsewhere
F5	420–512	Incredible: total demolition

The Fujita scale, named after tornado expert Theodore Fujita, ranks tornadoes according to how much damage they cause. Fortunately, only one percent of tornadoes are classified as violent (F4 and F5).

Global Warming Some scientists hypothesize that Earth's ocean temperatures are increasing due to global warming. In your Science Journal, predict what might happen to the strength of hurricanes if Earth's oceans become warmer.

Figure 15 In this hurricane cross section, the small, red arrows indicate rising, warm, moist air. This air forms cumulus and cumulonimbus clouds in bands around the eye. The green arrows indicate cool, dry air sinking in the eye and between the cloud bands.

Hurricanes The most powerful storm is the hurricane. A **hurricane,** illustrated in **Figure 15,** is a large, swirling, low-pressure system that forms over the warm Atlantic Ocean. It is like a machine that turns heat energy from the ocean into wind. A storm must have winds of at least 119 km/h to be called a hurricane. Similar storms are called typhoons in the Pacific Ocean and cyclones in the Indian Ocean.

Hurricanes are similar to low-pressure systems on land, but they are much stronger. In the Atlantic and Pacific Oceans, low pressure sometimes develops near the equator. In the northern hemisphere, winds around this low pressure begin rotating counterclockwise. The strongest hurricanes affecting North America usually begin as a low-pressure system west of Africa. Steered by surface winds, these storms can travel west, gaining strength from the heat and moisture of warm ocean water.

When a hurricane strikes land, high winds, tornadoes, heavy rains, and high waves can cause a lot of damage. Floods from the heavy rains can cause additional damage. Hurricane weather can destroy crops, demolish buildings, and kill people and other animals. As long as a hurricane is over water, the warm, moist air rises and provides energy for the storm. When a hurricane reaches land, however, its supply of energy disappears and the storm loses power.

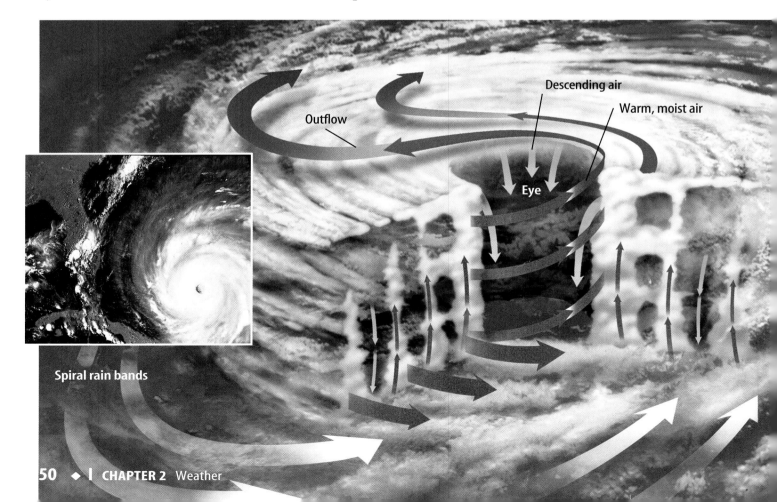

Descending air

Warm, moist air

Outflow

Eye

Spiral rain bands

Blizzards Severe storms also can occur in winter. If you live in the northern United States, you may have awakened from a winter night's sleep to a cold, howling wind and blowing snow, like the storm in **Figure 16.** The National Weather Service classifies a winter storm as a **blizzard** if the winds are 56 km/h, the temperature is low, the visibility is less than 400 m in falling or blowing snow, and if these conditions persist for three hours or more.

Severe Weather Safety When severe weather threatens, the National Weather Service issues a watch or warning. Watches are issued when conditions are favorable for severe thunderstorms, tornadoes, floods, blizzards, and hurricanes. During a watch, stay tuned to a radio or television station reporting the weather. When a warning is issued, severe weather conditions already exist. You should take immediate action. During a severe thunderstorm or tornado warning, take shelter in the basement or a room in the middle of the house away from windows. When a hurricane or flood watch is issued, be prepared to leave your home and move farther inland.

Blizzards can be blinding and have dangerously low temperatures with high winds. During a blizzard, stay indoors. Spending too much time outside can result in severe frostbite.

Figure 16 Blizzards can be extremely dangerous because of their high winds, low temperatures, and poor visibility.

---

section **2** review

Summary

Weather Changes
- Air masses tend to have temperature and moisture properties similar to Earth's surface.
- Winds blow from areas of high pressure to areas of lower pressure.

Fronts
- A boundary between different air masses is called a front.

Severe Weather
- The National Weather Service issues watches or warnings, depending on the severity of the storm, for people's safety.

Self Check

1. **Draw Conclusions** Why is fair weather common during periods of high pressure?
2. **Describe** how a cold front affects weather.
3. **Explain** what causes lightning and thunder.
4. **Compare and contrast** a watch and a warning. How can you keep safe during a tornado warning?
5. **Think Critically** Explain why some fronts produce stronger storms than others.

Applying Skills

6. **Recognize Cause and Effect** Describe how an occluded front may form over your city and what effects it can have on the weather.

Weather Forecasts

What You'll Learn

- **Explain** how data are collected for weather maps and forecasts.
- **Identify** the symbols used in a weather station model.

Why It's Important

Weather observations help you predict future weather events.

⏺ Review Vocabulary

forecast: to predict a condition or event on the basis of observations

New Vocabulary

- meteorologist
- isotherm
- station model
- isobar

Figure 17 A meteorologist uses Doppler radar to track a tornado. Since the nineteenth century, technology has greatly improved weather forecasting.

Weather Observations

You can determine current weather conditions by checking the thermometer and looking to see whether clouds are in the sky. You know when it's raining. You have a general idea of the weather because you are familiar with the typical weather where you live. If you live in Florida, you don't expect snow in the forecast. If you live in Maine, you assume it will snow every winter. What weather concerns do you have in your region?

A **meteorologist** (mee tee uh RAH luh jist) is a person who studies the weather. Meteorologists take measurements of temperature, air pressure, winds, humidity, and precipitation. Computers, weather satellites, Doppler radar shown in **Figure 17,** and instruments attached to balloons are used to gather data. Such instruments improve meteorologists' ability to predict the weather. Meteorologists use the information provided by weather instruments to make weather maps. These maps are used to make weather forecasts.

Forecasting Weather

Meteorologists gather information about current weather and use computers to make predictions about future weather patterns. Because storms can be dangerous, you do not want to be unprepared for threatening weather. However, meteorologists cannot always predict the weather exactly because conditions can change rapidly.

The National Weather Service depends on two sources for its information—data collected from the upper atmosphere and data collected on Earth's surface. Meteorologists of the National Weather Service collect information recorded by satellites, instruments attached to weather balloons, and from radar. This information is used to describe weather conditions in the atmosphere above Earth's surface.

Station Models When meteorologists gather data from Earth's surface, it is recorded on a map using a combination of symbols, forming a **station model.** A station model, like the one in **Figure 18,** shows the weather conditions at a specific location on Earth's surface. Information provided by station models and instruments in the upper atmosphere is entered into computers and used to forecast weather.

Temperature and Pressure In addition to station models, weather maps have lines that connect locations of equal temperature or pressure. A line that connects points of equal temperature is called an **isotherm** (I suh thurm). *Iso* means "same" and *therm* means "temperature." You probably have seen isotherms on weather maps on TV or in the newspaper.

An **isobar** is a line drawn to connect points of equal atmospheric pressure. You can tell how fast wind is blowing in an area by noting how closely isobars are spaced. Isobars that are close together indicate a large pressure difference over a small area. A large pressure difference causes strong winds. Isobars that are spread apart indicate a smaller difference in pressure. Winds in this area are gentler. Isobars also indicate the locations of high- and low-pressure areas.

Reading Check *How do isobars indicate wind speed?*

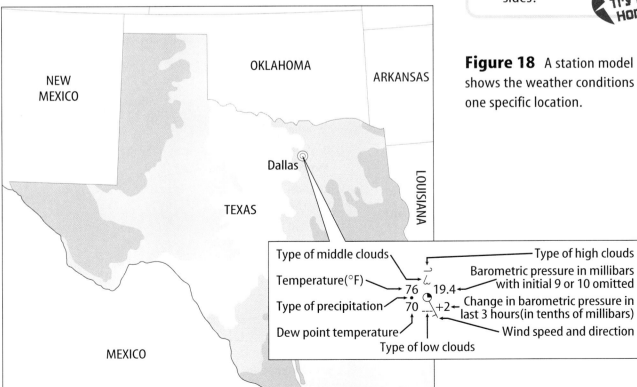

Figure 18 A station model shows the weather conditions at one specific location.

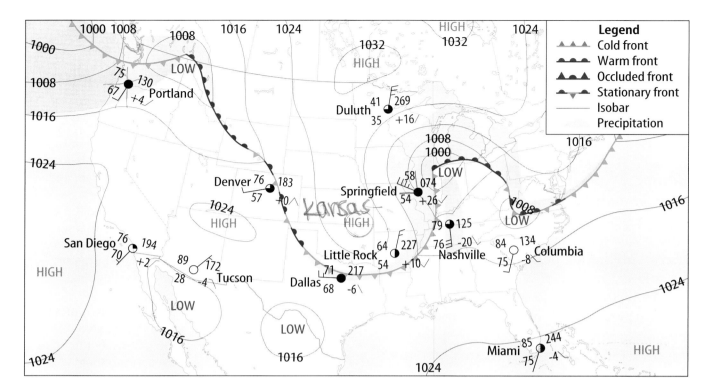

Figure 19 Highs, lows, isobars, and fronts on this weather map help meteorologists forecast the weather.

Weather Maps On a weather map like the one in **Figure 19,** pressure areas are drawn as circles with the word High or Low in the middle of the circle. Fronts are drawn as lines and symbols. When you watch weather forecasts on television, notice how weather fronts move from west to east. This is a pattern that meteorologists depend on to forecast weather.

section 3 review

Summary

Weather Observations

- Meteorologists are people who study the weather and make weather maps.

Forecasting Weather

- Meteorologists gather information about current weather and make predictions about future weather patterns.

- A station model shows weather conditions at a specific location on Earth's surface by using symbols to record meteorological data.

- On weather maps, isotherms are lines that connect points of equal temperature.

- An isobar is a line drawn on a weather map that connects points of equal atmospheric pressure.

Self Check

1. **List** some instruments that are used to collect weather data.

2. **Describe** at least six items of data that might be recorded in a station model.

3. **Explain** how the National Weather Service makes weather maps.

4. **Explain** what closely spaced isobars on a weather map indicate.

5. **Think Critically** In the morning you hear a meteorologist forecast today's weather as sunny and warm. After school, it is raining. Why is the weather so hard to predict?

Applying Skills

6. **Concept Map** Using a computer, make an events-chain concept map for how a weather forecast is made.

Reading a Weather Map

Meteorologists use a series of symbols to provide a picture of local and national weather conditions. With what you know, can you interpret weather information from weather map symbols?

◉ Real-World Question

How do you read a weather map?

Materials
magnifying lens
Weather Map Symbols Appendix
Figure 19 (Weather Map)

Goals
- ■ **Learn** how to read a weather map.
- ■ **Use** information from a station model and a weather map to forecast weather.

◉ Procedure

Use the information provided in the questions below and the Weather Map Symbols Appendix to learn how to read a weather map.

1. Find the station models on the map for Portland, Oregon, and Miami, Florida. Find the dew point, wind direction, barometric pressure, and temperature at each location.

2. Looking at the placement of the isobars, determine whether the wind would be stronger at Springfield, Illinois, or at San Diego, California. Record your answer. What is another way to determine the wind speed at these locations?

3. **Determine** the type of front near Dallas, Texas. Record your answer.

4. The triangles or half-circles are on the side of the line toward the direction the front is moving. In which direction is the cold front located over Washington state moving?

◉ Conclude and Apply

1. Locate the pressure system over southeast Kansas. Predict what will happen to the weather of Nashville, Tennessee, if this pressure system moves there.

2. Prevailing westerlies are winds responsible for the movement of much of the weather across the United States. Based on this, would you expect Columbia, South Carolina, to continue to have clear skies? Explain.

3. The direction line on the station model indicates the direction from which the wind blows. The wind is named for that direction. Infer from this the name of the wind blowing at Little Rock, Arkansas.

𝒞ommunicating Your Data

Pretend you are a meteorologist for a local TV news station. Make a poster of your weather data and present a weather forecast to your class.

Model and Invent

Measuring Wind Speed

Goals
- **Invent** an instrument or devise a system for measuring wind speeds using common materials.
- **Devise** a method for using your invention or system to compare different wind speeds.

Possible Materials
paper
scissors
confetti
grass clippings
meterstick
*measuring tape
*Alternate materials

Safety Precautions

Data Source
Refer to Section 1 for more information about anemometers and other wind speed instruments. Consult the data table for information about Beaufort's wind speed scale.

▶ Real-World Question

When you watch a gust of wind blow leaves down the street, do you wonder how fast the wind is moving? For centuries, people could only guess at wind speeds, but in 1805, Admiral Beaufort of the British navy invented a method for estimating wind speeds based on their effect on sails. Later, Beaufort's system was modified for use on land. Meteorologists use a simple instrument called an anemometer to measure wind speeds, and they still use Beaufort's system to estimate the speed of the wind. What type of instrument or system can you invent to measure wind speed? How could you use simple materials to invent an instrument or system for measuring wind speeds? What observations do you use to estimate the speed of the wind?

▶ Make a Model

1. Scan the list of possible materials and choose the materials you will need to devise your system.
2. **Devise** a system to measure different wind speeds. Be certain the materials you use are light enough to be moved by slight breezes.

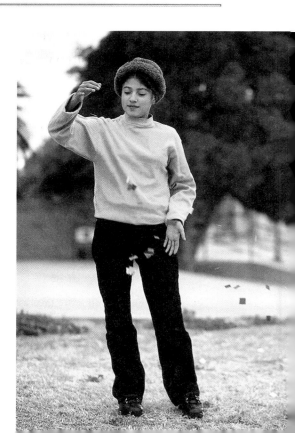

Check the Model Plans

1. **Describe** your plan to your teacher. Provide a sketch of your instrument or system and ask your teacher how you might improve its design.

2. Present your idea for measuring wind speed to the class in the form of a diagram or poster. Ask your classmates to suggest improvements in your design that will make your system more accurate or easier to use.

▶ Test Your Model

1. Confetti or grass clippings that are all the same size can be used to measure wind speed by dropping them from a specific height. Measuring the distances they travel in different strength winds will provide data for devising a wind speed scale.

2. Different sizes and shapes of paper also could be dropped into the wind, and the strength of the wind would be determined by measuring the distances traveled by these different types of paper.

▶ Analyze Your Data

1. **Develop** a scale for your method.

2. **Compare** your results with Beaufort's wind speed scale.

3. **Analyze** what problems may exist in the design of your system and suggest steps you could take to improve your design.

▶ Conclude and Apply

1. **Explain** why it is important for meteorologists to measure wind speeds.

2. **Evaluate** how well your system worked in gentle breezes and strong winds.

Beaufort's Wind Speed Scale	
Description	Wind Speed (km/h)
calm—smoke drifts up	less than 1
light air—smoke drifts with wind	1–5
light breeze—leaves rustle	6–11
gentle breeze—leaves move constantly	12–19
moderate breeze—branches move	20–29
fresh breeze—small trees sway	30–39
strong breeze—large branches move	40–50
moderate gale—whole trees move	51–61
fresh gale—twigs break	62–74
strong gale—slight damage to houses	75–87
whole gale—much damage to houses	88–101
storm—extensive damage	102–120
hurricane—extreme damage	more than 120

*C*ommunicating Your Data

Demonstrate your system for the class. Compare your results and measurements with the results of other classmates.

Rainmakers

Cloud seeding is an inexact science

You listen to a meteorologist give the long-term weather forecast. Another week with no rain in sight. As a farmer, you are concerned that your crops are withering in the fields. Home owners' lawns are turning brown. Wildfires are possible. Cattle are starving. And, if farmers' crops die, there could be a shortage of food and prices will go up for consumers.

Meanwhile, several states away, another farmer is listening to the weather report calling for another week of rain. Her crops are getting so water soaked that they are beginning to rot.

Weather. Can't scientists find a way to better control it? The answer is...not exactly. Scientists have been experimenting with methods to control our weather since the 1940s. And nothing really works.

Cloud seeding is one such attempt. It uses technology to enhance the natural rainfall process. The idea has been used to create rain where it is needed or to reduce hail damage. Government officials also use cloud seeding or weather modification to try to reduce the force of a severe storm.

Some people seed a cloud by flying a plane above it and releasing highway-type flares with chemicals, such as silver iodide. Another method is to fly beneath the cloud and spray a chemical that can be carried into the cloud by air currents.

Cloud seeding doesn't work with clouds that have little water vapor or are not near the dew point. Seeding chemicals must be released into potential rain clouds. The chemicals provide nuclei for water molecules to cluster around. Water then falls to Earth as precipitation.

Cloud seeding does have its critics. If you seed clouds and cause rain for your area, aren't you preventing rain from falling in another area? Would that be considered "rain theft" by people who live in places where the cloudburst would naturally occur? What about those cloud-seeding agents? Could the cloud-seeding chemicals, such as silver iodide and acetone, affect the environment in a harmful way? Are humans meddling with nature and creating problems in ways that haven't been determined?

Flares are lodged under a plane. The pilot will drop them into potential rain clouds.

Debate Learn more about cloud seeding and other methods of changing weather. Then debate whether or not cloud seeding can be considered "rain theft."

Science Online

For more information, visit booki.msscience.com/time

Reviewing Main Ideas

Section 1 What is weather?

1. Factors that determine weather include air pressure, wind, temperature, and the amount of moisture in the air.

2. More water vapor can be present in warm air than in cold air. Water vapor condenses when the dew point is reached. Clouds are formed when warm, moist air rises and cools to its dew point.

3. Rain, hail, sleet, and snow are types of precipitation.

Section 2 Weather Patterns

1. Fronts form when air masses with different characteristics meet. Types of fronts include cold, warm, occluded, and stationary fronts.

2. High atmospheric pressure at Earth's surface usually means good weather. Cloudy and stormy weather occurs under low pressure.

3. Tornadoes, thunderstorms, hurricanes, and blizzards are examples of severe weather.

Section 3 Weather Forecasts

1. Meteorologists use information from radar, satellites, computers, and other weather instruments to forecast the weather.

2. Weather maps include information about temperature and air pressure. Station models indicate weather at a particular location.

Visualizing Main Ideas

Copy and complete the following concept map about air temperature, water vapor, and pressure.

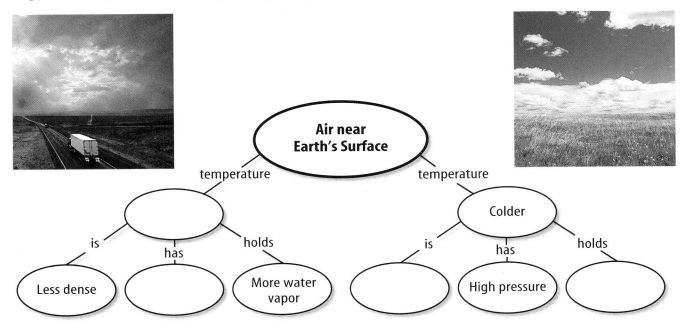

Using Vocabulary

air mass p. 44
blizzard p. 51
dew point p. 39
fog p. 41
front p. 45
humidity p. 38
hurricane p. 50
isobar p. 53

isotherm p. 53
meteorologist p. 52
precipitation p. 42
relative humidity p. 38
station model p. 53
tornado p. 48
weather p. 36

Explain the differences between the vocabulary words in each of the following sets.

1. air mass—front

2. humidity—relative humidity

3. relative humidity—dew point

4. dew point—precipitation

5. hurricane—tornado

6. blizzard—fog

7. meteorologist—station model

8. precipitation—fog

9. isobar—isotherm

10. isobar—front

Checking Concepts

Choose the word or phrase that best answers the question.

11. Which term refers to the amount of water vapor in the air?
    A) dew point   **C)** humidity
    B) precipitation **D)** relative humidity

12. What does an anemometer measure?
    A) wind speed   **C)** air pressure
    B) precipitation **D)** relative humidity

13. Which type of air has a relative humidity of 100 percent?
    A) humid   **C)** dry
    B) temperate   **D)** saturated

Use the photo below to answer question 14.

14. Which type of the following clouds are high feathery clouds made of ice crystals?
    A) cirrus   **C)** cumulus
    B) nimbus   **D)** stratus

15. What is a large body of air that has the same properties as the area over which it formed called?
    A) air mass   **C)** front
    B) station model **D)** isotherm

16. At what temperature does water vapor in air condense?
    A) dew point   **C)** front
    B) station model **D)** isobar

17. Which type of precipitation forms when water vapor changes directly into a solid?
    A) rain   **C)** sleet
    B) hail   **D)** snow

18. Which type of front may form when cool air, cold air, and warm air meet?
    A) warm   **C)** stationary
    B) cold   **D)** occluded

19. Which is issued when severe weather conditions exist and immediate action should be taken?
    A) front   **C)** station model
    B) watch   **D)** warning

20. What is a large, swirling storm that forms over warm, tropical water called?
    A) hurricane   **C)** blizzard
    B) tornado   **D)** hailstorm

Thinking Critically

21. **Explain** the relationship between temperature and relative humidity.

22. **Describe** how air, water, and the Sun interact to cause weather.

23. **Explain** why northwest Washington often has rainy weather and southwest Texas is dry.

24. **Determine** What does it mean if the relative humidity is 79 percent?

25. **Infer** Why don't hurricanes form in Earth's polar regions?

26. **Compare and contrast** the weather at a cold front and the weather at a warm front.

27. **Interpret Scientific Illustrations** Use the cloud descriptions in this chapter to describe the weather at your location today. Then try to predict tomorrow's weather.

28. **Compare and contrast** tornadoes and thunderstorms. Include information about wind location and direction.

29. **Concept Map** Copy and complete the sequence map below showing how precipitation forms.

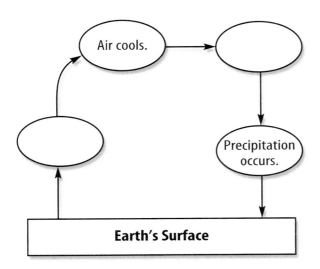

Air cools.

Precipitation occurs.

Earth's Surface

30. **Observe and Infer** You take a hot shower. The mirror in the bathroom fogs up. Infer from this information what has happened.

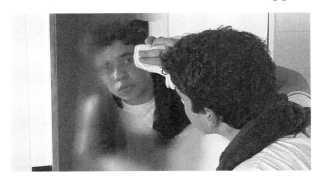

Performance Activities

31. **Board Game** Make a board game using weather terms. You could make cards to advance or retreat a token.

32. **Design** your own weather station. Record temperature, precipitation, and wind speed for one week.

Applying Math

Use the table below to answer question 33.

Air Temperature (°C)	Amount of Water Vapor Needed for Saturation (g/m³)
25	22
20	15

33. **Dew Point** If the air temperature is 25°C and the relative humidity is 55 percent, will the dew point be reached if the temperature drops to 20°C?

34. **Rising Temperature** If the air temperature is 30°C and the relative humidity is 60 percent, will the dew point be reached if the temperature rises to 35°C? Use the graph in **Figure 4** to explain your answer.

Part 1 | Multiple Choice

Record your answers on the answer sheet provided by your teacher or on a sheet of paper.

Use the table and paragraph below to answer questions 1 and 2.

Hurricanes are rated on a scale based on their wind speed and barometric pressure. The table below lists hurricane categories.

Hurricane Rating Scale		
Category	Wind Speed (km/h)	Barometric Pressure (millibars)
1	119–154	>980
2	155–178	965–980
3	179–210	945–964
4	211–250	920–944
5	>250	<920

1. Hurricane Mitch, with winds of 313 km/h and a pressure of 907 mb, struck the east coast of Central America in 1998. What category was Hurricane Mitch?
   A. 2　　　　C. 4
   B. 3　　　　D. 5

2. Which of the following is true when categorizing a hurricane?
   A. Storm category increases as wind increases and pressure decreases.
   B. Storm category increases as wind decreases and pressure increases.
   C. Storm category increases as wind and pressure increase.
   D. Storm category decreases as wind and pressure decrease.

Test-Taking Tip

Fill In All Blanks Never leave any answer blank.

3. Which of the following instruments is used to measure air pressure?
   A. anemometer　　C. barometer
   B. thermometer　　D. rain gauge

4. Which of the following is a description of a tornado?
   A. a large, swirling, low-pressure system that forms over the warm Atlantic Ocean
   B. a winter storm with winds at least 56 km/h and low visibility
   C. a violently rotating column of air in contact with the ground
   D. a boundary between two air masses of different density, moisture, or temperature

Use the figure below to answer questions 5 and 6.

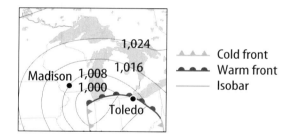

5. What is the atmospheric pressure in the city of Madison, Wisconsin?
   A. 1024 mb　　C. 1008 mb
   B. 1016 mb　　D. 1000 mb

6. What type of front is near Toledo, Ohio?
   A. cold front
   B. warm front
   C. stationary front
   D. occluded front

7. Which of the following terms is used to describe a person studies the weather?
   A. meteorologist
   B. geologist
   C. biologist
   D. paleontologist

Part 2 | Short Response/Grid In

Record your answer on the answer sheet provided by your teacher or on a sheet of paper.

8. Compare and contrast the formation of stratus clouds, cumulus clouds and cirrus clouds.

Use the graph below to answer questions 9 and 10.

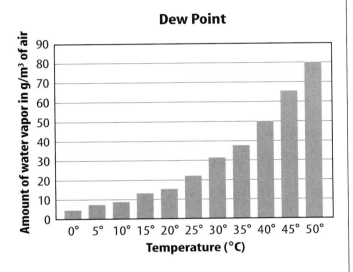

Dew Point

9. What amount of water vapor is in air at a temperature of 45°C?

10. On a fall day, the relative humidity is 72 percent and the temperature is 30°C. Will the dew point be reached if the temperature drops to 20°C? Why or why not?

11. Describe weather conditions during which hailstones form and the process by which they form.

12. What effects do high-pressure systems have on air circulation and weather? What effects do low-pressure systems have on weather?

13. Explain the relationship between differences in atmospheric pressure and wind speed.

Part 3 | Open Ended

Record your answers on a sheet of paper.

14. Explain the relationship between lightning and thunder.

15. Describe how a hurricane in the Northern Hemisphere forms.

16. Explain why hurricanes lose power once they reach land.

Use the figure below to answer question 17.

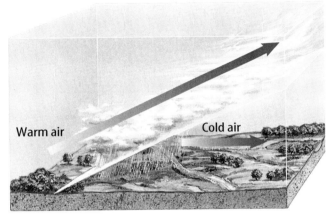

Warm air   Cold air

17. What type of front is shown? How does this type of front form?

18. Explain what type of weather occurs at front boundaries.

19. List the safety precautions you should take during a severe weather alert, including tornado warnings, flood watches, and blizzards, respectively.

20. Explain how the Sun's heat energy creates Earth's weather.

21. What are the four main types of precipitation? Describe the differences between each type.

Climate

The BIG Idea

Climate is the pattern of weather that occurs in an area over many years.

SECTION 1
What is climate?
Main Idea Some factors that affect the climate of a region include latitude, landforms, location of lakes and oceans, and ocean currents.

SECTION 2
Climate Types
Main Idea World climates can be classified by using averages of temperature and precipitation and the vegetation that is adapted to an area.

SECTION 3
Climate Changes
Main Idea The causes of climatic change can operate over short periods of time or very long periods of time.

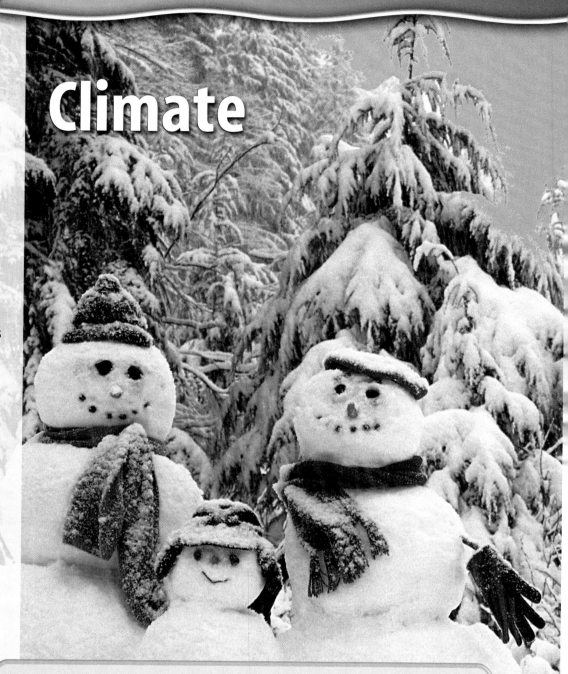

Why do seasons change?

Why do some places have four distinct seasons, while others have only a wet and dry season? In this chapter, you will learn what climate is and how climates are classified. You will also learn what causes climate changes and how humans and animals adapt to different climates.

Science Journal Write a paragraph explaining what you already know about the causes of seasons.

Start-Up Activities

Tracking World Climates

You wouldn't go to Alaska to swim or to Jamaica to snow ski. You know the climates in these places aren't suited for these sports. In this lab, you'll explore the climates in different parts of the world.

1. Obtain a world atlas, globe, or large classroom map. Select several cities from different parts of the world.

2. Record the longitude and latitude of your cities. Note if they are near mountains or an ocean.

3. Research the average temperature of your cities. In what months are they hottest? Coldest? What is the average yearly rainfall? What kinds of plants and animals live in the region? Record your findings.

4. Compare your findings with those of the rest of your class. Can you see any relationship between latitude and climate? Do cities near an ocean or a mountain range have different climatic characteristics?

5. **Think Critically** Keep track of the daily weather conditions in your cities. Are these representative of the kind of climates your cities are supposed to have? Suggest reasons why day-to-day weather conditions may vary.

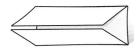

Classifying Climates Make the following Foldable to help you compare climatic types.

STEP 1 Fold two pieces of paper lengthwise into thirds.

STEP 2 Fold the papers widthwise into fourths.

STEP 3 Unfold, lay the papers lengthwise, and draw lines along the folds as shown.

STEP 4 Label your tables as shown.

Climate Classification		
Tropical		
Mild		
Dry		

Climate Classification		
Continental		
Polar		
High elevation		

Make a Table As you read the chapter, define each type of climate and write notes on its weather characteristics.

Preview this chapter's content and activities at booki.msscience.com

Get Ready to Read

Monitor

1 Learn It! An important strategy to help you improve your reading is monitoring, or finding your reading strengths and weaknesses. As you read, monitor yourself to make sure the text makes sense. Discover different monitoring techniques you can use at different times, depending on the type of test and situation.

2 Practice It! The paragraph below appears in Section 2. Read the passage and answer the questions that follow. Discuss your answers with other students to see how they monitor their reading.

> Climatologists—people who study climates— usually use a system developed in 1918 by Wladimir Köppen to classify climates. Köppen observed that the types of plants found in a region depended on the climate of the area.... He classified world climates by using the annual and monthly averages of temperature and precipitation of different regions. He then related the types and distribution of native vegetation to the various climates.
>
> *—from page 70*

- What questions do you still have after reading?
- Do you understand all of the words in the passage?
- Did you have to stop reading often? Is the reading level appropriate for you?

3 Apply It! Identify one paragraph that is difficult to understand. Discuss it with a partner to improve your understanding.

Reading Tip

Monitor your reading by slowing down or speeding up depending on your understanding of the text.

Target Your Reading

Use this to focus on the main ideas as you read the chapter.

1. **Before you read** the chapter, respond to the statements below on your worksheet or on a numbered sheet of paper.
   - Write an **A** if you **agree** with the statement.
   - Write a **D** if you **disagree** with the statement.

2. **After you read** the chapter, look back to this page to see if you've changed your mind about any of the statements.
   - If any of your answers changed, explain why.
   - Change any false statements into true statements.
   - Use your revised statements as a study guide.

Science Online

Print out a worksheet of this page at booki.msscience.com

Before You Read A or D	Statement	After You Read A or D
	1 Climate is determined by averaging the weather of a region over a long period of time.	
	2 Latitude affects the climate of an area.	
	3 Ocean currents do not affect weather and climate.	
	4 The area on the leeward (downwind) side of a mountain experiences high rainfall.	
	5 There is only one type of climate in North America.	
	6 Over the past 100 years, Earth's average global surface temperature has increased.	
	7 El Niño can affect weather patterns, leading to droughts in some areas and flooding in others.	
	8 During the past century, atmospheric carbon dioxide has decreased.	
	9 Deforestation affects the amount of carbon dioxide in the atmosphere.	

What is climate?

What You'll Learn

■ **Describe** what determines climate.
■ **Explain** how latitude, oceans, and other factors affect the climate of a region.

Why It's Important

Climate affects the way you live.

🔍 **Review Vocabulary**
latitudes: distance in degrees north or south of the equator

New Vocabulary
● climate   ● polar zone
● tropics   ● temperate zone

Figure 1 The tropics are warmer than the temperate zones and the polar zones because the tropics receive the most direct solar energy.

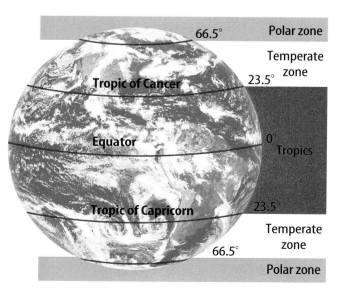

Climate

If you wandered through a tropical rain forest, you would see beautiful plants flowering in shades of pink and purple beneath a canopy of towering trees. A variety of exotic birds and other animals would dart among the tree branches and across the forest floor. The sounds of singing birds and croaking frogs would surround you. All of these organisms thrive in hot temperatures and abundant rainfall. Rain forests have a hot, wet climate. **Climate** is the pattern of weather that occurs in an area over many years. It determines the types of plants or animals that can survive, and it influences how people live.

Climate is determined by averaging the weather of a region over a long period of time, such as 30 years. Scientists average temperature, precipitation, air pressure, humidity, and number of days of sunshine to determine an area's climate. Some factors that affect the climate of a region include latitude, landforms, location of lakes and oceans, and ocean currents.

Latitude and Climate

As you can see in **Figure 1,** regions close to the equator receive the most solar radiation. Latitude, a measure of distance north or south of the equator, affects climate. **Figure 2** compares cities at different latitudes. The **tropics**—the region between latitudes 23.5°N and 23.5°S—receive the most solar radiation because the Sun shines almost directly over these areas. The tropics have temperatures that are always hot, except at high elevations. The **polar zones** extend from 66.5°N and 66.5°S latitude to the poles. Solar radiation hits these zones at a low angle, spreading energy over a large area. During winter, polar regions receive little or no solar radiation. Polar regions are never warm.

✔ **Reading Check** *How does latitude affect climate?*

Between the tropics and the polar zones are the **temperate zones.** Temperatures here are moderate. Most of the United States is in a temperate zone.

Other Factors

In addition to the general climate divisions of polar, temperate, and tropical, natural features such as large bodies of water, ocean currents, and mountains affect climate within each zone. Large cities also change weather patterns and influence the local climate.

Large Bodies of Water If you live or have vacationed near an ocean, you may have noticed that water heats up and cools down more slowly than land does. This is because it takes a lot more heat to increase the temperature of water than it takes to increase the temperature of land. In addition, water must give up more heat than land does for it to cool. Large bodies of water can affect the climate of coastal areas by absorbing or giving off heat. This causes many coastal regions to be warmer in the winter and cooler in the summer than inland areas at similar latitude. Look at **Figure 2** again. You can see the effect of an ocean on climate by comparing the average temperatures in a coastal city and an inland city, both located at 37°N latitude.

Figure 2 This map shows average daily low temperatures in four cities during January and July. It also shows average yearly precipitation.

Mini LAB

Observing Solar Radiation

Procedure
1. Darken the room.
2. Hold a **flashlight** about 30 cm from a **globe.** Shine the light directly on the equator. With your finger, trace around the light.
3. Now, tilt the flashlight to shine on 30°N latitude. The size of the lighted area should increase. Repeat at 60°N latitude.

Analysis
1. How did the size and shape of the light beam change as you directed the light toward higher latitudes?
2. How does Earth's tilt affect the solar radiation received by different latitudes?

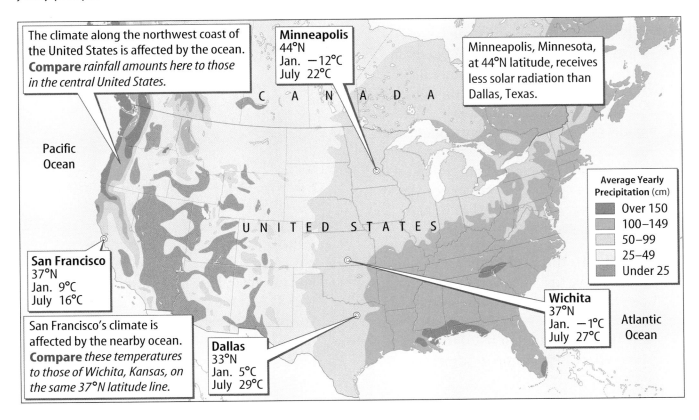

The climate along the northwest coast of the United States is affected by the ocean. **Compare** *rainfall amounts here to those in the central United States.*

Pacific Ocean

Minneapolis
44°N
Jan. −12°C
July 22°C

C A N A D A

Minneapolis, Minnesota, at 44°N latitude, receives less solar radiation than Dallas, Texas.

U N I T E D   S T A T E S

San Francisco
37°N
Jan. 9°C
July 16°C

San Francisco's climate is affected by the nearby ocean. **Compare** *these temperatures to those of Wichita, Kansas, on the same 37°N latitude line.*

Dallas
33°N
Jan. 5°C
July 29°C

Wichita
37°N
Jan. −1°C
July 27°C

Atlantic Ocean

Average Yearly Precipitation (cm)
- Over 150
- 100–149
- 50–99
- 25–49
- Under 25

Mountain Air When air rises over a mountain, the air expands and its temperature decreases, causing water vapor to condense and form rain. Temperature changes caused by air expanding or contracting also occur in some machines. Why does the air coming out of a bicycle pump feel cold?

Ocean Currents Ocean currents affect coastal climates. Warm currents begin near the equator and flow toward higher latitudes, warming the land regions they pass. When the currents cool off and flow back toward the equator, they cool the air and climates of nearby land.

✔ **Reading Check** *How do ocean currents affect climate?*

Winds blowing from the sea are often moister than those blowing from land. Therefore, some coastal areas have wetter climates than places farther inland. Look at the northwest coast of the United States shown in **Figure 2.** The large amounts of precipitation in Washington, Oregon, and northern California can be explained by this moist ocean air.

Mountains At the same latitude, the climate is colder in the mountains than at sea level. When radiation from the Sun is absorbed by Earth's surface, it heats the land. Heat from Earth then warms the atmosphere. Because Earth's atmosphere gets thinner at higher altitudes, the air in the mountains has fewer molecules to absorb heat.

Applying Science

How do cities influence temperature?

The temperature in a city can be several degrees warmer than the temperature of nearby rural areas. This difference in temperature is called the heat island effect. Cities contain asphalt and concrete which heat up rapidly as they absorb energy from the Sun. Rural areas covered with vegetation stay cooler because plants and soil contain water. Water heats up more slowly and carries away heat as it evaporates. Is the heat island effect the same in summer and winter?

Identifying the Problem

The table lists the average summer and winter high temperatures in and around a city in 1996 and 1997. By examining the data, can you tell if the heat island effect is the same in summer and winter?

Average Seasonal Temperatures		
Season	Temperature (°C)	
	City	Rural
Winter 1996	−3.0	−4.4
Summer 1996	23.5	20.9
Winter 1997	−0.1	−1.8
Summer 1997	23.6	21.2

Solving the Problem

1. Calculate the average difference between city and rural temperatures in summer and in winter. In which season is the heat island effect the largest?
2. For this area there are about 15 hours of daylight in summer and 9 hours in winter. Use this fact to explain your results from the previous question.

Rain Shadows Mountains also affect regional climates, as shown in **Figure 3.** On the windward side of a mountain range, air rises, cools, and drops its moisture. On the leeward side of a mountain range air descends, heats up, and dries the land. Deserts are common on the leeward sides of mountains.

Cities Large cities affect local climates. Streets, parking lots, and buildings heat up, in turn heating the air. Air pollution traps this heat, creating what is known as the heat-island effect. Temperatures in a city can be 5°C higher than in surrounding rural areas.

Figure 3 Large mountain ranges can affect climate by forcing air to rise over the windward side, cooling and bringing precipitation. The air descends with little or no moisture, creating desertlike conditions on the leeward side.

section 1 review

Summary

Latitude and Climate

- Climate is the pattern of weather that occurs in an area over many years.
- The tropics receive the most solar radiation because the Sun shines most directly there.
- The polar zones receive the least solar energy due to the low-angled rays.
- Temperate zones, located between the tropics and the polar zones, have moderate temperatures.

Other Factors

- Natural features such as large bodies of water, ocean currents, and mountains can affect local and regional climates.
- Large cities can change weather patterns and influence local climates.

Self Check

1. **Explain** how two cities located at the same latitude can have different climates.
2. **Describe** how mountains affect climate.
3. **Define** the heat island effect.
4. **Compare and contrast** tropical and polar climates.
5. **Think Critically** Explain why plants found at different elevations on a mountain might differ. How can latitude affect the elevation at which some plants are found?

Applying Math

6. **Solve One-Step Equations** The coolest average summer temperature in the United States is 2°C at Barrow, Alaska, and the warmest is 37°C at Death Valley, California. Calculate the range of average summer temperatures in the United States.

Climate Types

as you read

What **You'll Learn**

- **Describe** a climate classification system.
- **Explain** how organisms adapt to particular climates.

Why **It's Important**

Many organisms can survive only in climates to which they are adapted.

Review Vocabulary
regions: places united by specific characteristics

New Vocabulary
- adaptation
- hibernation

Figure 4 The type of vegetation in a region depends on the climate. **Describe** *what these plants tell you about the climate shown here.*

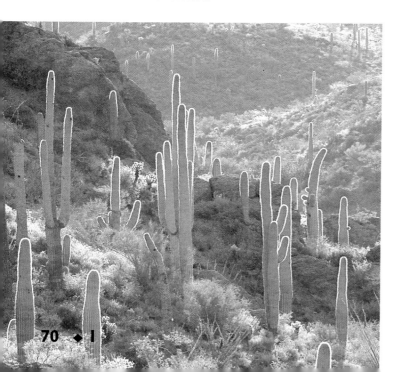

Classifying Climates

What is the climate like where you live? Would you call it generally warm? Usually wet and cold? Or different depending on the time of year? How would you classify the climate in your region? Life is full of familiar classification systems—from musical categories to food groups. Classifications help to organize your thoughts and to make your life more efficient. That's why Earth's climates also are classified and are organized into the various types that exist. Climatologists—people who study climates—usually use a system developed in 1918 by Wladimir Köppen to classify climates. Köppen observed that the types of plants found in a region depended on the climate of the area. **Figure 4** shows one type of region Köppen might have observed. He classified world climates by using the annual and monthly averages of temperature and precipitation of different regions. He then related the types and distribution of native vegetation to the various climates.

The climate classification system shown in **Figure 5** separates climates into six groups—tropical, mild, dry, continental, polar, and high elevation. These groups are further separated into types. For example, the dry climate classification is separated into semiarid and arid.

Adaptations

Climates vary around the world, and as Köppen observed, the type of climate that exists in an area determines the vegetation found there. Fir trees aren't found in deserts, nor are cacti found in rain forests. In fact, all organisms are best suited for certain climates. Organisms are adapted to their environment. An **adaptation** is any structure or behavior that helps an organism survive in its environment. Structural adaptations are inherited. They develop in a population over a long period of time. Once adapted to a particular climate, organisms may not be able to survive in other climates.

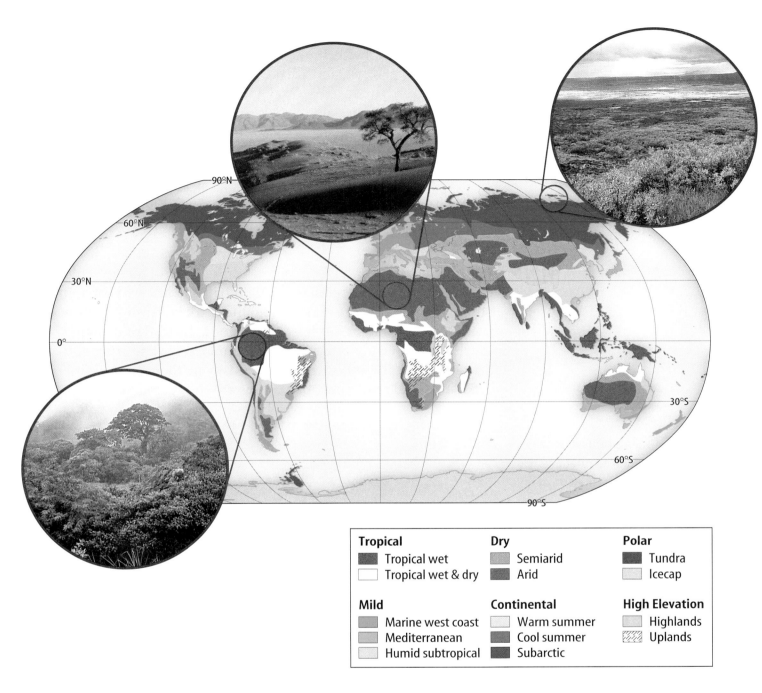

Tropical
- ■ Tropical wet
- □ Tropical wet & dry

Dry
- ▨ Semiarid
- ■ Arid

Polar
- ■ Tundra
- □ Icecap

Mild
- ▨ Marine west coast
- ▨ Mediterranean
- □ Humid subtropical

Continental
- □ Warm summer
- ■ Cool summer
- ■ Subarctic

High Elevation
- ▨ Highlands
- ▨ Uplands

Figure 5 This map shows a climate classification system similar to the one developed by Köppen. **Describe** *the patterns you can see in the locations of certain climate types.*

INTEGRATE
Life Science

Structural Adaptations Some organisms have body structures that help them survive in certain climates. The fur of mammals is really hair that insulates them from cold temperatures. A cactus has a thick, fleshy stem. This structural adaptation helps a cactus hold water. The waxy stem covering prevents water inside the cactus from evaporating. Instead of broad leaves, these plants have spiny leaves, called needles, that further reduce water loss.

✔ Reading Check *How do cacti conserve water?*

Behavioral Adaptations Some organisms display behavioral adaptations that help them survive in a particular climate. For example, rodents and certain other mammals undergo a period of greatly reduced activity in winter called **hibernation.** During hibernation, body temperature drops and body processes are reduced to a minimum. Some of the factors thought to trigger hibernation include cooler temperatures, shorter days, and lack of adequate food. The length of time that an animal hibernates varies depending on the particular species of animal and the environmental conditions.

✔ Reading Check *What is hibernation?*

Other animals have adapted differently. During cold weather, bees cluster together in a tight ball to conserve heat. On hot, sunny days, desert snakes hide under rocks. At night when it's cooler, they slither out in search of food. Instead of drinking water as turtles and lizards do in wet climates, desert turtles and lizards obtain the moisture they need from their food. Some behavioral and structural adaptations are shown in **Figure 6.**

Figure 6 Organisms have structural and behavioral adaptations that help them survive in particular climates.

The needles and the waxy skin of a cactus are structural adaptations to a desert climate. **Infer** *how these adaptations help cacti conserve water.*

These hibernating bats have adapted their behavior to survive winter.

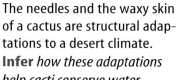

Polar bears have structural adaptations to keep them warm. The hairs of their fur trap air and heat.

Figure 7 Lungfish survive periods of intense heat and drought by going into an inactive state called estivation. During the dry season when water evaporates, lungfish dig into the mud and curl up in a small chamber they make at the lake's bottom. During the wet season, lungfish reemerge to live in small lakes and pools.

Estivation Lungfish, shown in **Figure 7,** survive periods of intense heat by entering an inactive state called estivation (es tuh VAY shun). As the weather gets hot and water evaporates, the fish burrows into mud and covers itself in a leathery mixture of mud and mucus. It lives this way until the warm, dry months pass.

Like other organisms, you have adaptations that help you adjust to climate. In hot weather, your sweat glands release water onto your skin. The water evaporates, taking some heat with it. As a result, you become cooler. In cold weather, you may shiver to help your body stay warm. When you shiver, the rapid muscle movements produce some heat. What other adaptations to climate do people have?

section 2 review

Summary

Classifying Climates

- Climatologists classify climates into six main groups: tropical, mild, dry, continental, polar, and high elevation.

Adaptations

- Adaptations are any structures or behaviors that help an organism to survive.

- Structural adaptations such as fur, hair, and spiny needles help an organism to survive in certain climates.

- Behavioral adaptations include hibernation, a period of greatly reduced activity in winter; estivation, an inactive state during intense heat; clustering together in the cold; and obtaining water from food when water is not found elsewhere.

Self Check

1. **List** Use **Figure 5** and a world map to identify the climate type for each of the following locations: Cuba, North Korea, Egypt, and Uruguay.

2. **Compare and contrast** hibernation and estivation.

3. **Think Critically** What adaptations help dogs keep cool during hot weather?

Applying Skills

4. **Form Hypotheses** Some scientists have suggested that Earth's climate is getting warmer. What effects might this have on vegetation and animal life in various parts of the United States?

5. **Communicate** Research the types of vegetation found in the six climate regions shown in **Figure 5.** Write a paragraph in your Science Journal describing why vegetation can be used to help define climate boundaries.

Climatic Changes

as you read

What You'll Learn

- **Explain** what causes seasons.
- **Describe** how El Niño affects climate.
- **Explore** possible causes of climatic change.

Why It's Important

Changing climates could affect sea level and life on Earth.

⊙ **Review Vocabulary**

solar radiation: energy from the Sun transferred by waves or rays

New Vocabulary

- season
- El Niño
- greenhouse effect
- global warming
- deforestation

Earth's Seasons

In temperate zones, you can play softball under the summer Sun and in the winter go sledding with friends. Weather changes with the season. **Seasons** are short periods of climatic change caused by changes in the amount of solar radiation an area receives. **Figure 8** shows Earth revolving around the Sun. Because Earth is tilted, different areas of Earth receive changing amounts of solar radiation throughout the year.

Seasonal Changes Because of fairly constant solar radiation near the equator, the tropics do not have much seasonal temperature change. However, they do experience dry and rainy seasons. The middle latitudes, or temperate zones, have warm summers and cool winters. Spring and fall are usually mild.

✔ **Reading Check** *What are seasons like in the tropics?*

Figure 8 As Earth revolves around the Sun, different areas of Earth are tilted toward the Sun, which causes different seasons.
Identify *During which northern hemisphere season is Earth closer to the Sun?*

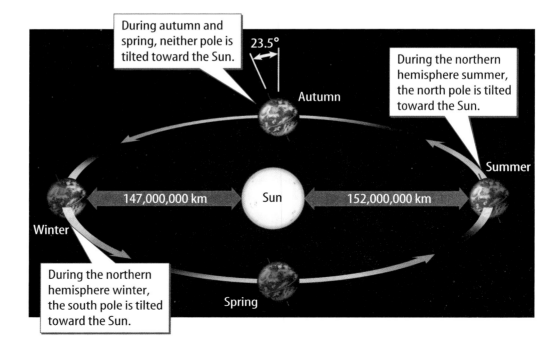

During autumn and spring, neither pole is tilted toward the Sun.

23.5°

During the northern hemisphere summer, the north pole is tilted toward the Sun.

Autumn

Summer

147,000,000 km   Sun   152,000,000 km

Winter

Spring

During the northern hemisphere winter, the south pole is tilted toward the Sun.

Figure 9 A strong El Niño, like the one that occurred in 1998, can affect weather patterns around the world.

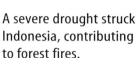
A severe drought struck Indonesia, contributing to forest fires.

California was plagued by large storms that produced pounding surf and shoreline erosion.

High Latitudes During the year, the high latitudes near the poles have great differences in temperature and number of daylight hours. As shown in **Figure 8,** during summer in the northern hemisphere, the north pole is tilted toward the Sun. During summer at the north pole, the Sun doesn't set for nearly six months. During that same time, the Sun never rises at the south pole. At the equator days are about the same length all year long.

El Niño and La Niña

El Niño (el NEEN yoh) is a climatic event that involves the tropical Pacific Ocean and the atmosphere. During normal years, strong trade winds that blow east to west along the equator push warm surface water toward the western Pacific Ocean. Cold, deep water then is forced up from below along the coast of South America. During El Niño years, these winds weaken and sometimes reverse. The change in the winds allows warm, tropical water in the upper layers of the Pacific to flow back eastward to South America. Cold, deep water is no longer forced up from below. Ocean temperatures increase by 1°C to 7°C off the coast of Peru.

El Niño can affect weather patterns. It can alter the position and strength of one of the jet streams. This changes the atmospheric pressure off California and wind and precipitation patterns around the world. This can cause drought in Australia and Africa. This also affects monsoon rains in Indonesia and causes storms in California, as shown in **Figure 9.**

The opposite of El Niño is La Niña, shown in **Figure 10.** During La Niña, the winds blowing across the Pacific are stronger than normal, causing warm water to accumulate in the western Pacific. The water in the eastern Pacific near Peru is cooler than normal. La Niña may cause droughts in the southern United States and excess rainfall in the northwestern United States.

Mini LAB

Modeling El Niño

Procedure

1. During El Niño, trade winds blowing across the Pacific Ocean from east to west slacken or even reverse. Surface waters move back toward the coast of Peru.
2. Add **warm water** to a **9-in × 13-in baking pan** until it is two-thirds full. Place the pan on a smooth countertop.
3. Blow as hard as you can across the surface of the water along the length of the pan. Next, blow with less force. Then, blow in the opposite direction.

Analysis

1. What happened to the water as you blew across its surface? What was different when you blew with less force and when you blew from the opposite direction?
2. Explain how this is similar to what happens during an El Niño event.

Try at Home

Figure 10

Weather in the United States can be affected by changes that occur thousands of kilometers away. Out in the middle of the Pacific Ocean, periodic warming and cooling of a huge mass of seawater—phenomena known as El Niño and La Niña, respectively—can impact weather across North America. During normal years (right), when neither El Niño nor La Niña is in effect, strong winds tend to keep warm surface waters contained in the western Pacific while cooler water wells up to the surface in the eastern Pacific.

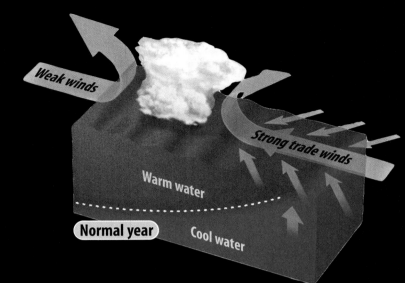

Weak winds
Strong trade winds
Warm water
Normal year
Cool water

EL NIÑO During El Niño years, winds blowing west weaken and may even reverse. When this happens, warm waters in the western Pacific move eastward, preventing cold water from upwelling. These changes can alter global weather patterns and trigger heavier-than-normal precipitation across much of the United States.

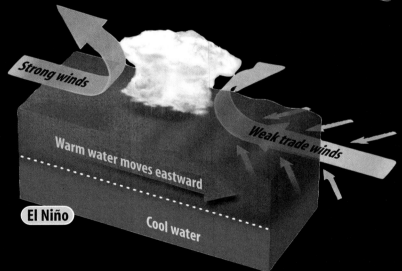

Strong winds
Weak trade winds
Warm water moves eastward
El Niño
Cool water

Very weak winds
Very strong trade winds
Warm water moves westward
La Niña
Cool water

LA NIÑA During La Niña years, stronger-than-normal winds push warm Pacific waters farther west, toward Asia. Cold, deep-sea waters then well up strongly in the eastern Pacific, bringing cooler and often drier weather to many parts of the United States.

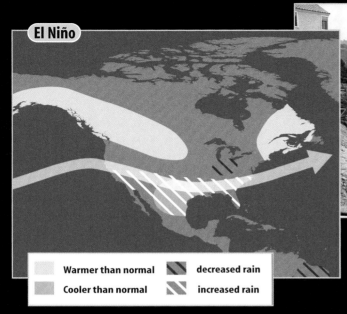

El Niño

Warmer than normal   decreased rain

Cooler than normal   increased rain

Sun-warmed surface water spans the Pacific Ocean during El Niño years. Clouds form above the warm ocean, carrying moisture aloft. The jet stream, shown by the white arrow above, helps bring some of this warm, moist air to the United States.

▲ **LANDSLIDE** Heavy rains in California resulting from El Niño can lead to landslides. This upended house in Laguna Niguel, California, took a ride downhill during the El Niño storms of 1998.

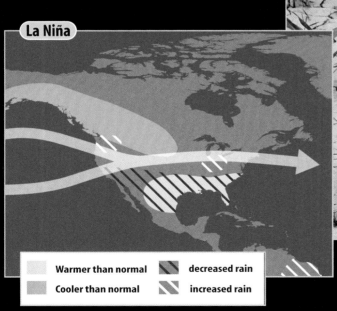

La Niña

Warmer than normal   decreased rain

Cooler than normal   increased rain

During a typical La Niña year, warm ocean waters, clouds, and moisture are pushed away from North America. A weaker jet stream often brings cooler weather to the northern parts of the continent and hot, dry weather to southern areas.

▲ **PARCHED LAND** The Southeast may experience drought conditions, like those that struck the cornfields of Montgomery County, Maryland, during the La Niña summer of 1988.

Climatic Change

If you were exploring in Antarctica near Earth's south pole and found a 3-million-year-old fossil of a warm-weather plant or animal, what would it tell you? You might conclude that the climate of that region changed because Antarctica is much too cold for similar plants and animals to survive today. Some warm-weather fossils found in polar regions indicate that at times in Earth's past, worldwide climate was much warmer than at present. At other times Earth's climate has been much colder than it is today.

Sediments in many parts of the world show that at several different times in the past 2 million years, glaciers covered large parts of Earth's surface. These times are called ice ages. During the past 2 million years, ice ages have alternated with warm periods called interglacial intervals. Ice ages seem to last 60,000 to 100,000 years. Most interglacial periods are shorter, lasting 10,000 to 15,000 years. We are now in an interglacial interval that began about 11,500 years ago. Additional evidence suggests that climate can change even more quickly. Ice cores record climate in a way similar to tree rings. Cores drilled in Greenland show that during the last ice age, colder times lasting 1,000 to 2,000 years changed quickly to warmer spells that lasted about as long. **Figure 11** shows a scientist working with ice cores.

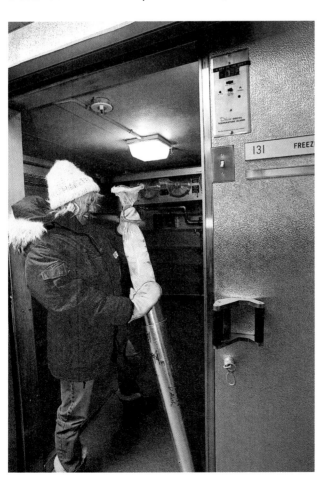

Figure 11 Some ice cores consist of layers of ice that record detailed climate information for individual years. These ice cores can cover more than 300,000 years. **Describe** *how this is helpful.*

What causes climatic change?

Climatic change has many varied causes. These causes of climatic change can operate over short periods of time or very long periods of time. Catastrophic events, including meteorite collisions and large volcanic eruptions, can affect climate over short periods of time, such as a year or several years. These events add solid particles and liquid droplets to the upper atmosphere, which can change climate. Another factor that can alter Earth's climate is short- or long-term changes in solar output, which is the amount of energy given off by the Sun. Changes in Earth's movements in space affect climate over many thousands of years, and movement of Earth's crustal plates can change climate over millions of years. All of these things can work separately or together to alter Earth's climate.

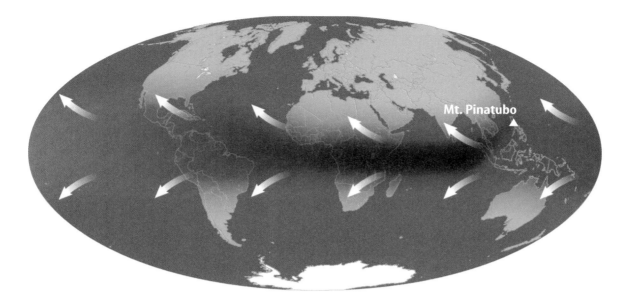

Atmospheric Solids and Liquids Small solid and liquid particles always are present in Earth's atmosphere. These particles can enter the atmosphere naturally or be added to the atmosphere by humans as pollution. Some ways that particles enter the atmosphere naturally include volcanic eruptions, soot from fires, and wind erosion of soil particles. Humans add particles to the atmosphere through automobile exhaust and smokestack emissions. These small particles can affect climate.

Catastrophic events such as meteorite collisions and volcanic eruptions put enormous volumes of dust, ash, and other particles into the atmosphere. These particles block so much solar radiation that they can cool the planet. **Figure 12** shows how a major volcanic eruption affected Earth's atmosphere.

In cities, particles put into the atmosphere as pollution can change the local climate. These particles can increase the amount of cloud cover downwind from the city. Some studies have even suggested that rainfall amounts can be reduced in these areas. This may happen because many small cloud droplets form rather than larger droplets that could produce rain.

Energy from the Sun Solar radiation provides Earth's energy. If the output of radiation from the Sun varies, Earth's climate could change. Some changes in the amount of energy given off by the Sun seem to be related to the presence of sunspots. Sunspots are dark spots on the surface of the Sun. **WARNING:** *Never look directly at the Sun.* Evidence supporting the link between sunspots and climate includes an extremely cold period in Europe between 1645 and 1715. During this time, very few sunspots appeared on the Sun.

Figure 12 Mount Pinatubo in the Philippines erupted in 1991. During the eruption, particles were spread high into the atmosphere and circled the globe. Over time, particles spread around the world, blocking some of the Sun's energy from reaching Earth. The gray areas show how particles from the eruption moved around the world.

Air Quality Control/Monitor
Atmospheric particles from pollution can affect human health as well as climate. These small particles, often called particulates, can enter the lungs and cause tissue damage. The Department of Environmental Protection employs people to monitor air pollution and its causes. Research what types of laws air quality control monitors must enforce.

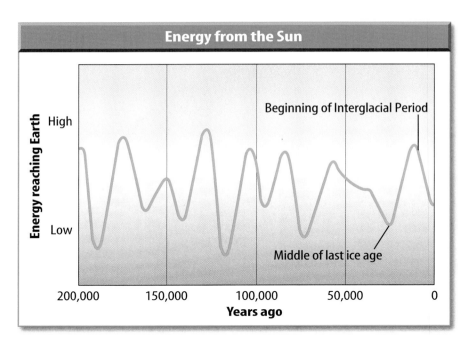

Energy from the Sun

Beginning of Interglacial Period

Middle of last ice age

Energy reaching Earth — High / Low

200,000   150,000   100,000   50,000   0

Years ago

Figure 13 The curving line shows how the amount of the Sun's energy that strikes the northern hemisphere changed over the last 200,000 years.
Describe *the amount of energy that reached the northern hemisphere during the last ice age.*

Earth Movements Another explanation for some climatic changes involves Earth's movements in space. Earth's axis currently is tilted 23.5° from perpendicular to the plane of its orbit around the Sun. In the past, this tilt has increased to 24.5° and has decreased to 21.5°. When this tilt is at its maximum, the change between summer and winter is probably greater. Earth's tilt changes about every 41,000 years. Some scientists hypothesize that the change in tilt affects climate.

Two additional Earth movements also cause climatic change. Earth's axis wobbles in space just like the axis of a top wobbles when it begins to spin more slowly. This can affect the amount of solar energy received by different parts of Earth. Also, the shape of Earth's orbit changes. Sometimes it is more circular than at present and sometimes it is more flattened. The shape of Earth's orbit changes over a 100,000-year cycle.

Amount of Solar Energy These movements of Earth cause the amount of solar energy reaching different parts of Earth to vary over time, as shown in **Figure 13.** These changes might have caused glaciers to grow and shrink over the last few million years. However, they do not explain why glaciers have occurred so rarely over long spans of geologic time.

Crustal Plate Movement Another explanation for major climatic change over tens or hundreds of millions of years concerns the movement of Earth's crustal plates. The movement of continents and oceans affects the transfer of heat on Earth, which in turn affects wind and precipitation patterns. Through time, these altered patterns can change climate. One example of this is when movement of Earth's plates created the Himalaya about 40 million years ago. The growth of these mountains changed climate over much of Earth.

As you've learned, many theories attempt to answer questions about why Earth's climate has changed through the ages. Probably all of these things play some role in changing climates. More study needs to be done before all the factors that affect climate will be understood.

Climatic Changes Today

Beginning in 1992, representatives from many countries have met to discuss the greenhouse effect and global climate change. These subjects also have appeared frequently in the headlines of newspapers and magazines. Some people are concerned that the greenhouse effect could be responsible for some present-day warming of Earth's atmosphere and oceans.

The **greenhouse effect** is a natural heating process that occurs when certain gases in Earth's atmosphere trap heat. Radiation from the Sun strikes Earth's surface and causes it to warm. Some of this heat then is radiated back toward space. Some gases in the atmosphere, known as greenhouse gases, absorb a portion of this heat and then radiate heat back toward Earth, as shown in **Figure 14.** This keeps Earth warmer than it would be otherwise.

There are many natural greenhouse gases in Earth's atmosphere. Water vapor, carbon dioxide, and methane are some of the most important ones. Without these greenhouse gases, life would not be possible on Earth. Like Mars, Earth would be too cold. However, if the greenhouse effect is too strong, Earth could get too warm. High levels of carbon dioxide in its atmosphere indicate that this has happened on the planet Venus.

Science nline

Topic: Greenhouse Effect
Visit booki.msscience.com for Web links to information about the greenhouse effect.

Activity Research changes in the greenhouse effect over the last 200 years. Infer what might be causing the changes.

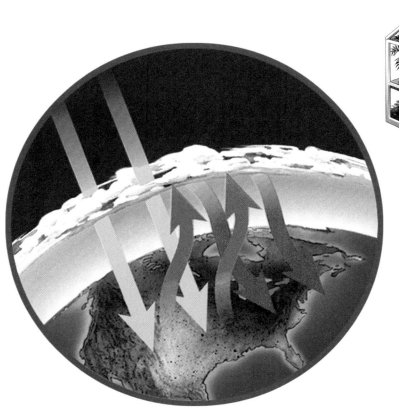

Figure 14 The Sun's radiation travels through Earth's atmosphere and heats the surface. Gases in our atmosphere trap the heat.
Compare and contrast *this to the way a greenhouse works.*

Global Warming

Over the past 100 years, the average global surface temperature on Earth has increased by about 0.6°C. This increase in temperature is known as **global warming.** Over the same time period, atmospheric carbon dioxide has increased by about 20 percent. As a result, researchers hypothesize that the increase in global temperatures may be related to the increase in atmospheric carbon dioxide. Other hypotheses include the possibility that global warming might be caused by changes in the energy emitted by the Sun.

If Earth's average temperature continues to rise, many glaciers could melt. When glaciers melt, the extra water causes sea levels to rise. Low-lying coastal areas could experience increased flooding. Already some ice caps and small glaciers are beginning to melt and recede, as shown in **Figure 15.** Sea level is rising in some places. Some scientific studies show that these events are related to Earth's increased temperature.

You learned in the previous section that organisms are adapted to their environments. When environments change, can organisms cope? In some tropical waters around the world, corals are dying. Many people think these deaths are caused by warmer water to which the corals are not adapted.

Some climate models show that in the future, Earth's temperatures will increase faster than they have in the last 100 years. However, these predictions might change because of uncertainties in the climate models and in estimating future increases in atmospheric carbon dioxide.

Figure 15 This glacier in Greenland might have receded from its previous position because of global warming. The pile of sediment in front shows how far the glacier once reached.

Figure 16 When forests are cleared or burned, carbon dioxide levels increase in the atmosphere.

Human Activities

Human activities affect the air in Earth's atmosphere. Burning fossil fuels and removing vegetation increase the amount of carbon dioxide in the atmosphere. Because carbon dioxide is a greenhouse gas, it might contribute to global warming. Each year, the amount of carbon dioxide in the atmosphere continues to increase.

Burning Fossil Fuels When natural gas, oil, and coal are burned for energy, the carbon in these fossil fuels combines with atmospheric oxygen to form carbon dioxide. This increases the amount of carbon dioxide in Earth's atmosphere. Studies indicate that humans have increased carbon dioxide levels in the atmosphere by about 25 percent over the last 150 years.

Deforestation Destroying and cutting down trees, called **deforestation**, also affects the amount of carbon dioxide in the atmosphere. Forests, such as the one shown in **Figure 16,** are cleared for mining, roads, buildings, and grazing cattle. Large tracts of forest have been cleared in every country on Earth. Tropical forests have been decreasing at a rate of about one percent each year for the past two decades.

As trees grow, they take in carbon dioxide from the atmosphere. Trees use this carbon dioxide to produce wood and leaves. When trees are cut down, the carbon dioxide they could have removed from the atmosphere remains in the atmosphere. Cut-down trees often are burned for fuel or to clear the land. Burning trees produces even more carbon dioxide.

Science⊙nline

Topic: Deforestation
Visit booki.msscience.com for Web links to information about deforestation.

Activity Collect data on the world's decline in forests. Infer what the world's forests will be like in 100 years.

 **Reading Check** *What can humans do to slow carbon dioxide increases in the atmosphere?*

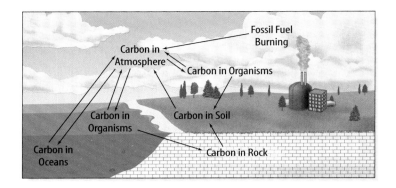

Figure 17 Carbon constantly is cycled among the atmosphere, oceans, solid earth, and biosphere.

The Carbon Cycle

Carbon, primarily as carbon dioxide, is constantly recycled in nature among the atmosphere, Earth's oceans, and organisms that inhabit the land. Organisms that undergo photosynthesis on land and in the water take in carbon dioxide and produce and store carbon-based food. This food is consumed by non-photosynthetic organisms. Carbon dioxide is released as food is broken down to release energy. When organisms die and decay, some carbon is stored as humus in soil and some carbon is released as carbon dioxide. This carbon cycle is illustrated in **Figure 17.**

Some carbon dioxide in the atmosphere dissolves in the oceans, and is used by algae and other photosynthetic, aquatic organisms. Just as on land, aquatic organisms give off carbon dioxide. However, Earth's oceans currently absorb more carbon dioxide from the atmosphere than they give off.

When Earth's climate changes, the amount of carbon dioxide that cycles among atmosphere, ocean, and land also can change. Some people hypothesize that if Earth's climate continues to warm, more carbon dioxide may be absorbed by oceans and land. Scientists continue to collect data to study any changes in the global carbon cycle.

section 3 review

Summary

Earth's Seasons

- Seasons are short periods of climatic changes due to Earth's tilt on its axis while revolving around the Sun, causing differing amounts of solar energy to reach areas of Earth.

El Niño and La Niña

- El Niño begins in the tropical Pacific Ocean when trade winds weaken or reverse directions, disrupting the normal temperature and precipitation patterns around the globe.

Climatic Changes Today

- The greenhouse effect is a natural heating process that occurs when certain gases in Earth's atmosphere trap heat.
- Burning fossil fuels increases the amount of carbon dioxide in the air.
- Deforestation increases the amount of carbon dioxide in the atmosphere.

Self Check

1. **Explain** how Earth's tilted axis is responsible for seasons.
2. **Compare and contrast** El Niño and La Niña. What climate changes do they demonstrate?
3. **List** factors that can cause Earth's climate to change.
4. **Explain** how people are adding carbon dioxide to the atmosphere.
5. **Think Critically** If Earth's climate continues to warm, how might your community be affected?

Applying Skills

6. **Use Models** Using a globe, model the three movements of Earth in space that can cause climatic change.
7. **Use a word processor** to make a table that lists the different processes that might cause Earth's climate to change. Include in your table a description of the process and how it causes climate to change.

The Greenhouse Effect

Do you remember climbing into the car on a warm, sunny day? Why was it so hot inside the car when it wasn't that hot outside? It was hotter in the car because the car functioned like a greenhouse. You experienced the greenhouse effect.

◉ Real-World Question

How can you demonstrate the greenhouse effect?

Goals
- **Model** the greenhouse effect.
- **Measure and graph** temperature changes.

Materials
identical large, empty glass jars (2)
lid for one jar
nonmercury thermometers (3)

Safety Precautions

WARNING: *Be careful when you handle glass thermometers. If a thermometer breaks, do not touch it. Have your teacher dispose of the glass safely.*

◉ Procedure

1. Lay a thermometer inside each jar.
2. Place the jars next to each other by a sunny window. Lay the third thermometer between the jars.
3. **Record** the temperatures of the three thermometers. They should be the same.
4. Place the lid on one jar.
5. **Record** the temperatures of all three thermometers at the end of 5, 10, and 15 min.
6. Make a line graph that shows the temperatures of the three thermometers for the 15 min of the experiment.

◉ Conclude and Apply

1. **Explain** why you placed a thermometer between the two jars.
2. **List** the constants in this experiment. What was the variable?
3. **Identify** which thermometer showed the greatest temperature change during your experiment.
4. **Analyze** what occurred in this experiment. How was the lid in this experiment like the greenhouse gases in the atmosphere?
5. **Infer** from this experiment why you should never leave a pet inside a closed car in warm weather.

Communicating Your Data

Give a brief speech describing your conclusions to your class.

MICROCLIMATES

Goals

- **Observe** temperature, wind speed, relative humidity, and precipitation in areas outside your school.
- **Identify** local microclimates.

Materials

thermometers
psychrometer
paper strip or wind sock
large cans (4 or 5)
* beakers or rain gauges (4 or 5)
unlined paper
*Alternate materials

Safety Precautions

WARNING: *If a thermometer breaks, do not touch it. Have your teacher dispose of the glass safely.*

Real-World Question

A microclimate is a localized climate that differs from the main climate of a region. Buildings in a city, for instance, can affect the climate of the surrounding area. Large buildings, such as the Bank of America Plaza in Dallas, Texas, can create microclimates by blocking the Sun or changing wind patterns. Does your school create microclimates?

Procedure

1. Select four or five sites around your school building. Also, select a control site well away from the school.

2. Attach a thermometer to an object near each of the locations you selected. Set up a rain gauge, beaker, or can to collect precipitation.

3. Visit each site at two predetermined times, one in the morning and one in the afternoon, each day for a week. Record the temperature and measure any precipitation that might have fallen. Use a wind sock or paper strip to determine wind direction.

Relative Humidity

Dry Bulb Temperature (°C)	Dry Bulb Temperature Minus Wet Bulb Temperature (°C)									
	1	2	3	4	5	6	7	8	9	10
14	90	79	70	60	51	42	34	26	18	10
15	90	80	71	61	53	44	36	27	20	13
16	90	81	71	63	54	46	38	30	23	15
17	90	81	72	64	55	47	40	32	25	18
18	91	82	73	65	57	49	41	34	27	20
19	91	82	74	65	58	50	43	36	29	22
20	91	83	74	66	59	51	44	37	31	24
21	91	83	75	67	60	53	46	39	32	26
22	92	83	76	68	61	54	47	40	34	28
23	92	84	76	69	62	55	48	42	36	30
24	92	84	77	69	62	56	49	43	37	31
25	92	84	77	70	63	57	50	44	39	33

4. To find relative humidity, you'll need to use a psychrometer. A psychrometer is an instrument with two thermometers—one wet and one dry. As moisture from the wet thermometer evaporates, it takes heat energy from its environment, and the environment immediately around the wet thermometer cools. The thermometer records a lower temperature. Relative humidity can be found by finding the difference between the wet thermometer and the dry thermometer and by using the chart on the previous page. Record all of your weather data.

◗ Analyze Your Data

1. Make separate line graphs for temperature, relative humidity, and precipitation for your morning and afternoon data. Make a table showing wind direction data.

2. **Compare and contrast** weather data for each of your sites. What microclimates did you identify around your school building? How did these climates differ from the control site? How did they differ from each other?

◗ Conclude and Apply

1. **Explain** Why did you take weather data at a control site away from the school building? How did the control help you analyze and interpret your data?

2. **Infer** what conditions could have caused the microclimates that you identified. Are your microclimates similar to those that might exist in a large city? Explain.

Communicating Your Data

Use your graphs to make a large poster explaining your conclusions. Display your poster in the school building. **For more help, refer to the** Science Skill Handbook.

The Year there was No Summer

You've seen pictures of erupting volcanoes. One kind of volcano sends smoke, rock, and ash high into the air above the crater. Another kind of volcano erupts with fiery, red-hot rivers of lava snaking down its sides. Erupting volcanoes are nature's forces at their mightiest, causing destruction and death. But not everyone realizes how far-reaching the destruction can be. Large volcanic eruptions can affect people thousands of kilometers away. In fact, major volcanic eruptions can have effects that reach around the globe.

An erupting volcano can temporarily change Earth's climate. The ash a volcano ejects into the atmosphere can create day after day without sunshine. Other particles move high into the atmosphere and are carried all the way around Earth, sometimes causing global temperatures to drop for several months.

The Summer That Never Came

An example of a volcanic eruption with wide-ranging effects occurred in 1783 in Iceland, an island nation in the North Atlantic Ocean. Winds carried a black cloud of ash from an erupting volcano in Iceland westward across northern Canada, Alaska, and across the Pacific Ocean to Japan. The summer turned bitterly cold in these places. Water froze, and heavy snowstorms pelted the land. Sulfurous gases from the erupting volcano combined with water to form particles of acid that reflected solar energy back into space. This "blanket" in the atmosphere kept the Sun's rays from heating up part of Earth.

The most tragic result of this eruption was the death of many Kauwerak people, who lived in western Alaska. Only a handful of Kauwerak survived the summer that never came. They had no opportunity to catch needed foods to keep them alive through the following winter.

Locate Using an atlas, locate Indonesia and Iceland. Using reference materials, find five facts about each place. Make a map of each nation and illustrate the map with your five facts.

Science Online

For more information, visit
booki.msscience.com/time

Reviewing Main Ideas

Section 1 | What is climate?

1. An area's climate is the average weather over a long period of time, such as 30 years.

2. The three main climate zones are tropical, polar, and temperate.

3. Features such as oceans, mountains, and even large cities affect climate.

Section 2 | Climate Types

1. Climates can be classified by various characteristics, such as temperature, precipitation, and vegetation. World climates commonly are separated into six major groups.

2. Organisms have structural and behavioral adaptations that help them survive in particular climates. Many organisms can survive only in the climate they are adapted to.

3. Adaptations develop in a population over a long period of time.

Section 3 | Climatic Changes

1. Seasons are caused by the tilt of Earth's axis as Earth revolves around the Sun.

2. El Niño disrupts normal temperature and precipitation patterns around the world.

3. Geological records show that over the past few million years, Earth's climate has alternated between ice ages and warmer periods.

4. The greenhouse effect occurs when certain gases trap heat in Earth's atmosphere.

5. Carbon dioxide enters the atmosphere when fossil fuels such as oil and coal are burned.

Visualizing Main Ideas

Copy and complete the following concept map on climate.

Climate

affected by — Latitude, ⬡, ⬡, Mountains

changes over — Short time [caused by Seasons, caused by Volcanoes, caused by ⬡], ⬡ [due to Earth movement, due to ⬡]

Using Vocabulary

adaptation p.70	hibernation p.72
climate p.66	polar zone p.66
deforestation p.83	season p.74
El Niño p.75	temperate zone p.66
global warming p.82	tropics p.66
greenhouse effect p.81	

Fill in the blanks with the correct vocabulary word or words.

1. Earth's north pole is in the _____.

2. _____ causes the Pacific Ocean to become warmer off the coast of Peru.

3. During _____, an animal's body temperature drops.

4. _____ is the pattern of weather that occurs over many years.

5. _____ means global temperatures are rising.

Checking Concepts

Choose the word or phrase that best answers the question.

6. Which of the following is a greenhouse gas in Earth's atmosphere?
   A) helium   **C)** hydrogen
   B) carbon dioxide   **D)** oxygen

7. During which of the following is the eastern Pacific warmer than normal?
   A) El Niño   **C)** summer
   B) La Niña   **D)** spring

8. Which latitude receives the most direct rays of the Sun year-round?
   A) 60°N   **C)** 30°S
   B) 90°N   **D)** 0°

9. What happens as you climb a mountain?
   A) temperature decreases
   B) temperature increases
   C) air pressure increases
   D) air pressure remains constant

10. Which of the following is true of El Niño?
    A) It cools the Pacific Ocean near Peru.
    B) It causes flooding in Australia.
    C) It cools the waters off Alaska.
    D) It may occur when the trade winds slacken or reverse.

11. What do changes in Earth's orbit affect?
    A) Earth's shape   **C)** Earth's rotation
    B) Earth's climate   **D)** Earth's tilt

12. The Köppen climate classification system includes categories based on precipitation and what other factor?
    A) temperature   **C)** winds
    B) air pressure   **D)** latitude

13. Which of the following is an example of structural adaptation?
    A) hibernation   **C)** fur
    B) migration   **D)** estivation

14. Which of these can people do in order to help reduce global warming?
    A) burn coal   **C)** conserve energy
    B) remove trees   **D)** produce methane

Use the illustration below to answer question 15.

N

15. What would you most likely find on the leeward side of this mountain range?
    A) lakes   **C)** deserts
    B) rain forests   **D)** glaciers

Thinking Critically

16. **Draw a Conclusion** How could climate change cause the types of organisms in an area to change?

17. **Infer** What might you infer if you find fossils of tropical plants in a desert?

18. **Describe** On a summer day, why would a Florida beach be cooler than an orange grove that is 2 km inland?

19. **Infer** what would happen to global climates if the Sun emitted more energy.

20. **Explain** why it will be cooler if you climb to a higher elevation in a desert.

21. **Communicate** Explain how atmospheric pressure over the Pacific Ocean might affect how the trade winds blow.

22. **Predict** Make a chain-of-events chart to explain the effect of a major volcanic eruption on climate.

23. **Form Hypotheses** A mountain glacier in South America has been getting smaller over several decades. What hypotheses should a scientist consider to explain why this is occurring?

24. **Concept Map** Copy and complete the concept map using the following: *tropics, 0°–23.5° latitude, polar, temperate, 23.5°–66.5° latitude,* and *66.5° latitude to poles.*

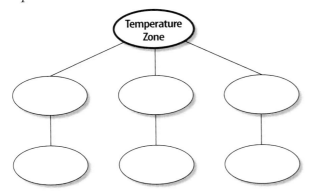

25. **Explain** how global warming might lead to the extinction of some organisms.

26. **Describe** how dust and ash from large volcanoes can change the atmosphere.

27. **Explain** how heat energy carried by ocean currents influences climate.

28. **Describe** how sediments, fossils, and ice cores record Earth's geologic history.

29. **Describe** how volcanic eruptions or meteorite collisions have changed past climates.

Performance Activities

30. **Science Display** Make a display illustrating different factors that can affect climate. Be sure to include detailed diagrams and descriptions for each factor in your display. Present your display to the class.

Applying Math

Use the table below to answer questions 31 and 32.

Precipitation in Phoenix, Arizona	
Season	Precipitation (cm)
Winter	5.7
Spring	1.2
Summer	6.7
Autumn	5.9
Total	19.5

31. **Precipitation Amounts** The following table gives average precipitation amounts for Phoenix, Arizona. Make a bar graph of these data. Which climate type do you think Phoenix represents?

32. **Local Precipitation** Use the table above to help estimate seasonal precipitation for your city or one that you choose. Create a bar graph for that data.

Part 1 Multiple Choice

Record your answers on the answer sheet provided by your teacher or on a sheet of paper.

Use the graph below to answer questions 1 and 2.

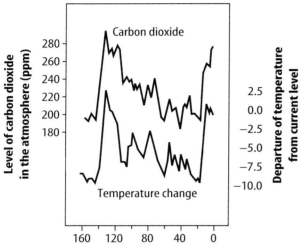

Years before present (in thousands)

1. Which of these statements is true according to the graph?
   A. Earth's mean temperature has never been hotter than it is today.
   B. The level of CO_2 has never been higher than today.
   C. The mean global temperature 60,000 years ago was less than today.
   D. The level of CO_2 in the atmosphere 80,000 years ago was 280 parts per million.

2. Which of the following statements best describes this graph?
   A. As CO_2 levels have increased, so has global temperature.
   B. As CO_2 levels have increased, global temperature has decreased.
   C. As global temperature has increased, CO_2 levels have decreased.
   D. No relationship exists between CO_2 and global temperatures.

Use the table below to answer question 3.

Apparent Temperature Index					
		Relative Humidity (%)			
		80	85	90	95
Air Temperature (°F)	85	97	99	102	105
	80	86	87	88	89
	75	78	78	79	79
	70	71	71	71	71

3. The National Weather Service created the Apparent Temperature Index to show the temperature the human body feels when heat and humidity are combined. If the relative humidity is 85% and the temperature is 75°F, what is the apparent temperature?
   A. 78°F   C. 88°F
   B. 79°F   D. 89°F

4. What is the most likely reason that the air temperature is warmest at the tropical latitudes?
   A. These latitudes receive the most solar radiation because there are no clouds.
   B. These latitudes receive the most solar radiation because the sun's angle is high.
   C. These latitudes receive the least solar radiation because the sun's angle is low.
   D. These latitudes receive the least solar radiation because of heavy cloud cover.

Test-Taking Tip

Qualifiers Look for quantifiers in a question. Such questions are not looking for absolute answers. Qualifiers would be words such as *most likely, most common,* or *least common.*

Question 2 Look for the most likely scientific explanation.

Part 2 | Short Response/Grid In

Record your answers on the answer sheet provided by your teacher or on a sheet of paper.

5. Explain how a large body of water can affect the climate of a nearby area.

6. Describe the relationship between ocean currents and precipitation in a coastal region.

7. The city of Redmond, Oregon is near the Cascade Mountain Range. The average annual rainfall for the Redmond, OR area is about 8 inches. Infer whether Redmond, OR is located on the windward side or leeward side of the mountain range. Explain your answer.

Use the figure below to answer question 8.

8. What is the greenhouse effect?

9. List three greenhouse gases.

10. How does the greenhouse effect positively affect life on Earth? How could it negatively affect life on Earth?

11. Explain why the temperature of a city can be up to 5°C warmer than the surrounding rural areas.

12. What are the different ways in which solid and liquid particles enter the Earth's atmosphere?

Part 3 | Open Ended

Record your answers on a sheet of paper.

Use the figure below to answer questions 13–15.

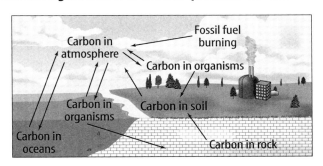

13. Describe the carbon cycle. Explain how carbon is transferred from organisms to soil.

14. How does the burning of fossil fuels affect the amount of carbon dioxide entering the carbon cycle?

15. How does deforestation affect the amount of carbon dioxide entering the carbon cycle?

16. In 1991 Mt. Pinatubo erupted, releasing volcanic particulates into the atmosphere. Temperatures around the world fell by as much as 0.7°C below average during 1992. How was this global temperature change related to the volcanic eruption?

17. What is global warming? What hypotheses help explain global warming? Explain the relationship between global warming and the level of seawater.

18. A scientist analyzes the pollen of ancient plants found preserved in lake sediments. The pollen is determined to be from a plant that needs moisture and year-round warm temperatures to grow. Make an inference about the type of climate that area experienced during the time the plant lived.

chapter 4

Air Pollution

The BIG Idea

People around the world live in areas with dangerously polluted air.

SECTION 1
Types and Causes of Air Pollution
Main Idea Air pollution comes from human activities as well as natural events.

SECTION 2
Effects of Air Pollution
Main Idea Air pollution not only affects your health and the health of other organisms, it also damages many materials.

SECTION 3
Solutions to Air Pollution
Main Idea Laws, conservation, and new technologies will help reduce air pollution.

Is the air always clear?

It's so sunny and clear that you can see the buildings of the city miles away. Some days, however, you notice a brown fog hanging in the air over the buildings. What causes this brown fog to form? Substances released into the air can affect your health and the health of other organisms.

Science Journal Write a paragraph describing three sources of air pollution.

Start-Up Activities

Acid in Air

Pollutants from power plants and car exhaust can react with water in the atmosphere to produce acids. In this lab, you will determine if your region receives acidic precipitation.

1. Place a clean, glass jar outside before a rain or snow.
2. Bring the jar indoors. If you collected snow, let it melt.
3. Use a piece of pH paper or a computer probe to test the pH of the water.
4. If the pH value you obtained was less than about 5.6, your region receives acid rain.
5. **Think Critically** Infer how air pollution in one region can cause acid rain hundreds of miles away.

Preview this chapter's content and activities at booki.msscience.com

FOLDABLES™
Study Organizer

Air Pollution Make the following Foldable to help you understand the vocabulary terms in this chapter.

STEP 1 Fold a vertical sheet of notebook paper from side to side.

STEP 2 Cut along every third line of only the top layer to form tabs.

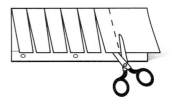

Build Vocabulary As you read the chapter, list the vocabulary words about air pollution on the tabs. As you learn the definitions, write them under the tab for each vocabulary word.

Get Ready to Read

Visualize

① Learn It! Visualize by forming mental images of the text as you read. Imagine how the text descriptions look, sound, feel, smell, or taste. Look for any pictures or diagrams on the page that may help you add to your understanding.

② Practice It! Read the following paragraph. As you read, use the underlined details to form a picture in your mind.

> The concentration of a pollutant in the air might not be high enough to cause a problem. Some pollutants, however, <u>stay in animal tissues</u> instead of being excreted from their bodies as waste. When these animals are <u>eaten by other animals</u>, the pollutants are <u>passed on to the predator</u>. **Biomagnification** (bi oh mag nuh fuh KAY shun) is the process in which <u>pollutant levels increase through the food chain</u>, as shown in **Figure 12.** Some fish are not safe for humans to eat frequently because of biomagnification.
>
> — *from page 107*

Based on the description above, try to visualize biomagnification. Now look at the illustration on page 107.
• How closely does it match your mental picture?
• Reread the passage and look at the picture again. Did your ideas change?
• Compare your image with what others in your class visualized.

③ Apply It! Read the chapter and list three subjects you were able to visualize. Make a rough sketch showing what you visualized.

Reading Tip

Forming your own mental images will help you remember what you read.

Target Your Reading

Use this to focus on the main ideas as you read the chapter.

1 **Before you read** the chapter, respond to the statements below on your worksheet or on a numbered sheet of paper.
- Write an **A** if you **agree** with the statement.
- Write a **D** if you **disagree** with the statement.

2 **After you read** the chapter, look back to this page to see if you've changed your mind about any of the statements.
- If any of your answers changed, explain why.
- Change any false statements into true statements.
- Use your revised statements as a study guide.

Science Online

Print out a worksheet of this page at booki.msscience.com

Before You Read A or D		Statement	After You Read A or D
	1	The term smog originally was used to describe the combination of smoke and fog.	
	2	Burning coal and oil releases pollutants that can combine with moisture in the air to form acid rain.	
	3	Chemicals called chlorofluorocarbons, or CFCs, are able to destroy ozone molecules.	
	4	Earth's ozone layer completely recovered in 1996 when industrialized nations stopped producing CFCs.	
	5	Children and elderly people experience the least effects of air pollution.	
	6	Acid from acid rain does not affect your lungs.	
	7	As Earth's ozone layer thins, Earth's organisms are exposed to more infrared radiation.	
	8	Increased exposure to ultraviolet radiation is linked to skin cancer and cataracts—a form of eye damage.	
	9	Laws and new technology will not affect air pollution.	

Types and Causes of Air Pollution

as you read

What You'll Learn

■ **Identify** the sources of air pollution.
■ **Describe** the effects of pollution on air quality.

Why It's Important

Understanding the causes of air pollution will help you learn ways to prevent it.

Review Vocabulary
pH scale: a scale used to measure how acidic or basic something is

New Vocabulary
● primary pollutant
● secondary pollutant
● photochemical smog
● acid rain
● particulate matter
● toxic air pollutant
● ozone layer

What causes air pollution?

Nearly every organism depends on gases like oxygen in Earth's atmosphere to carry out life functions. In addition to essential gases, the air you breathe also contains pollutants, which are harmful substances that contaminate the environment.

Air pollution comes from human activities as well as natural events. Industry, construction, power generation, transportation, and agriculture are a few examples of human activities that can pollute the air. Natural events that contribute to air pollution include erupting volcanoes that spew out ash and toxic gases. Kilauea Volcano on Hawaii is erupting and emitting toxic gases. People living downwind report having health problems. Smoke from forest fires and grass fires also can cause health problems.

Pollutants released directly into the air in a harmful form are called **primary pollutants.** Some examples of primary pollutants are shown in **Figure 1.** Pollutants that are not released directly into the air but form in the atmosphere are called **secondary pollutants.** Secondary pollutants are responsible for most of the brown haze, or smog, that you see near cities.

Figure 1 Primary pollutants include ash and toxic gases from volcanoes, soot from trucks, and smoke from industry smokestacks.

C Nitrogen dioxide and ozone form smog.

B Ultraviolet rays from the Sun help form ozone.

A Nitrogen compounds and organic compounds are released in car exhaust and form nitrogen dioxide.

Smog

The term *smog* originally was used to describe the combination of smoke and fog, but the smog you see near cities forms in a different way. Smog near cities is called **photochemical smog** because it forms with the help of sunlight. Photochemical smog forms when vehicles, some industries, and power plants release nitrogen compounds and organic compounds into the air. These substances react to form nitrogen dioxide. The nitrogen dioxide then can react in the presence of sunlight to eventually form ozone, a secondary pollutant, as shown in **Figure 2.** Ozone is a major component of smog, and nitrogen dioxide is a reddish-brown gas that contributes to the colored haze.

Reading Check *How does photochemical smog form?*

Nature and Smog Nature can affect the formation of smog. In many cities, smog is not a problem because winds disperse the pollutants that cause smog to form. In some locations, however, landforms can add to smog development. Los Angeles, California, for example, is a city that lies in a basin surrounded by the Santa Monica Mountains to the northwest, the San Gabriel Mountains to the north and east, and the Santa Ana Mountains to the southeast. These surrounding mountains trap air in the Los Angeles region, preventing pollutants from being dispersed quickly. Los Angeles also frequently has sunny, dry weather. When nitrogen compounds are added to the air and exposed to sunlight for long periods of time, thick blankets of smog can develop.

Figure 2 Pollutants from cars and other sources can cause urban smog.

City Smog Many large cities throughout the world have smog problems due to land forms, temperature inversions, population density and uncontrolled sources of pollution. Research how large cities in other countries, such as Mexico City, Mexico or Beijing, China protect people from smog.

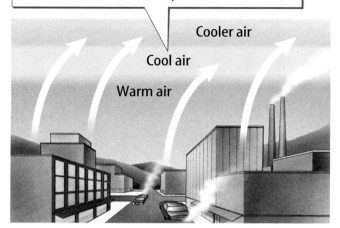

Cool air usually overlies warm air near Earth's surface. Pollutants can be carried away from their source.

Cooler air

Cool air

Warm air

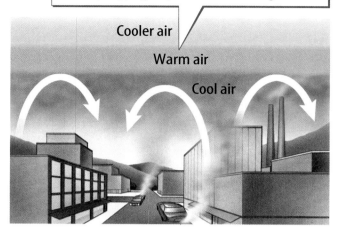

During a temperature inversion, warm air overlies cool air, trapping air pollutants near the ground.

Cooler air

Warm air

Cool air

Figure 3 A temperature inversion can worsen air pollution.

INTEGRATE Physics

Temperature Inversions The atmosphere also can influence the formation of smog, as shown in **Figure 3.** Normally, temperatures in Earth's lower atmosphere are warmest near Earth's surface. However, a temperature inversion sometimes occurs. During an inversion, warm air overlies cool air, trapping the cool air near Earth's surface. A temperature inversion reduces the amount of mixing in the atmosphere and can cause pollutants to accumulate near Earth's surface.

✔ Reading Check *What is a temperature inversion?*

Acid Rain

Acids and bases are two terms that describe substances. A substance that is neither acidic nor basic is neutral. The pH scale, shown in **Figure 4,** indicates how acidic or how basic a substance is. A pH of 7 is neutral. Substances with a pH lower than 7 are acids. Substances with a pH above 7 are bases. Rainwater is naturally slightly acidic, but pollution sometimes can cause rainwater to be even more acidic.

Natural lakes and streams have a pH between 6 and 8. **Acid rain** is precipitation with a pH below 5.6. When rain is acidic, the pH of lakes and streams may decrease.

Figure 4 Substances with a pH lower than 7 are acids. Those with a pH higher than 7 are bases. Rainwater naturally has a pH of about 5.6.

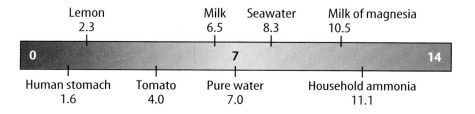

Lemon 2.3   Milk 6.5   Seawater 8.3   Milk of magnesia 10.5

0   7   14

Human stomach 1.6   Tomato 4.0   Pure water 7.0   Household ammonia 11.1

Acid Rain Sources Power plants burn fuels, like coal and oil, to produce the electricity that you need to light your home or power your stereo. Fuels also are burned for transportation and to heat your home. When fuels are burned, they release primary pollutants, such as sulfur dioxide and nitrogen oxides, into the air. These compounds rise into the atmosphere and combine with moisture in the air to form the secondary pollutants sulfuric and nitric acids.

Winds can carry acids long distances. The acids then can be returned to Earth's surface in precipitation. Acid rain can discolor painted surfaces, corrode metals, and damage concrete structures.

The Northeastern United States As shown in **Figure 5,** precipitation in the northeastern United States is more acidic than in other areas. Sulfur dioxides and nitrogen oxides released from midwestern power plants and other sources are carried by upper-level winds blowing from a generally westerly direction. The resulting acids that form in the atmosphere eventually return to Earth as acid rain. Soils and rocks are naturally more basic in the midwest than in the northeastern United States. Therefore, acid rain falling in New York can decrease the pH of soils, streams, or lakes more so than acid rain falling in Indiana. Many lakes in the northeastern United States have few fish due to acid rain.

Figure 5 The average pH of precipitation varies across the United States.

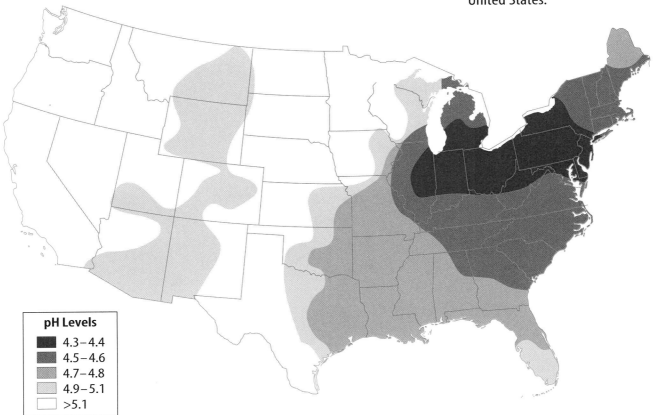

pH Levels
- 4.3–4.4
- 4.5–4.6
- 4.7–4.8
- 4.9–5.1
- >5.1

Observing Particulates

1. Spread **petroleum** jelly in a thin layer on a **small saucer.**
2. Place the saucer outside your home where it will not be disturbed.

Analysis
1. At the end of one day, examine the saucer.
2. Count the number of particulates you see in the petroleum jelly.
3. Infer where the particles might have come from.

Try at Home

Particulate Pollution

Air contains suspended solid particles and liquid droplets called **particulate matter.** Some particles enter the air directly and are therefore primary pollutants, such as smoke from a fireplace or soot in bus exhaust. Other particles, such as liquid droplets, can form from gases such as nitrogen or sulfur oxides as they combine with water in the air.

Coarse and Fine Particulates Coarse particulate matter is carried in the wind from dusty, unpaved roads, construction sites, and land that has been cleared. You can see coarse particulate matter easily when a lot of it is in the air, but the individual size of each particle is only about one-seventh the diameter of a human hair, as shown in **Figure 6.**

Fine particulate matter is much smaller than coarse particulate matter—only about one-fourth the size of coarse particulates. Fine particulates are released into the air from fires, vehicle exhaust, factories, and power plants. Particulate matter can damage plants and buildings and harm your lungs.

Figure 6 Coarse particulate matter is about one-seventh the diameter of a human hair. Fine particulates are much smaller.

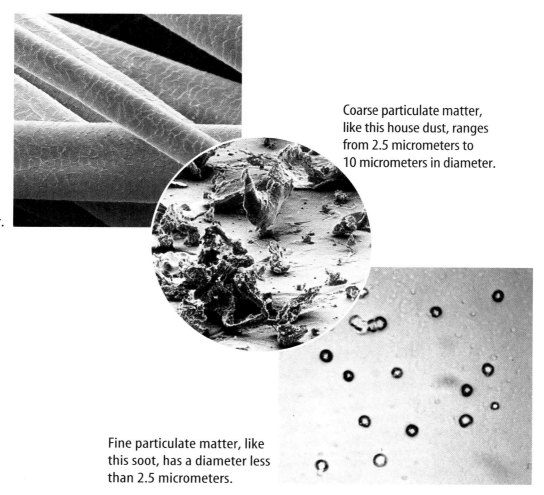

The average human hair is approximately 70 micrometers (0.000 07 m) in diameter.

Coarse particulate matter, like this house dust, ranges from 2.5 micrometers to 10 micrometers in diameter.

Fine particulate matter, like this soot, has a diameter less than 2.5 micrometers.

Figure 7 Toxic air pollution comes from factories, power plants, businesses, and transportation sources.

Mobile sources (50%) include cars, trucks, and planes.

Major sources (26%) include power plants and factories such as oil refineries and chemical manufacturers.

Area sources (24%) include businesses such as gas stations and dry cleaners.

Toxic Pollutants and Carbon Monoxide

More than 180 different substances released into the air are called **toxic air pollutants** because they cause or might cause cancer or other serious human health problems. Toxic pollutants also can damage other organisms. Most of the toxic air pollution is released by human activities, like those in **Figure 7.** Some air toxins can be released from natural events such as fires and erupting volcanoes.

When fossil fuels are not completely burned, a gas called carbon monoxide forms. In a typical U.S. city, as much as 95 percent of this colorless, odorless gas comes from mobile sources. Concentrations of carbon monoxide increase when cars are stopped in traffic. Low levels of carbon monoxide can harm people with heart disease. Carbon monoxide is poisonous at high concentrations.

Chlorofluorocarbons

Since their discovery in 1928, people have been using chemicals called chlorofluorocarbons (KLOR uh floor oh kar buhns), or CFCs, in air conditioners, refrigerators, and aerosol sprays. For many years, CFCs were thought to be wonder compounds. They don't burn. They're easy to manufacture. They aren't toxic. Millions of tons of CFCs were manufactured and sold by the mid-1970s. In 1974, scientists F. Sherwood Rowland and Mario Molina of the University of California began to wonder where all these CFCs ended up. They theorized that these compounds could end up high in Earth's atmosphere and damage Earth's ozone layer.

Science Online

Topic: Toxic Pollutants
Visit booki.msscience.com for Web links to information about toxic air pollution.

Activity List ten compounds considered to be toxic.

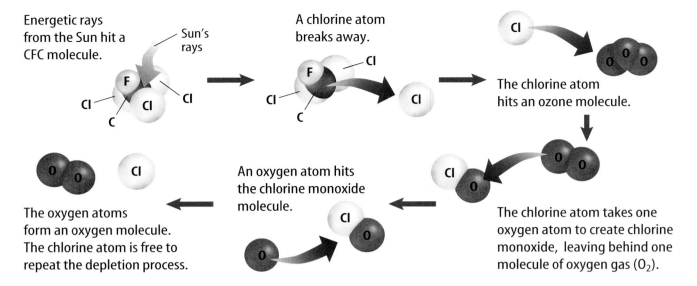

Energetic rays from the Sun hit a CFC molecule.

Sun's rays

A chlorine atom breaks away.

The chlorine atom hits an ozone molecule.

The chlorine atom takes one oxygen atom to create chlorine monoxide, leaving behind one molecule of oxygen gas (O_2).

An oxygen atom hits the chlorine monoxide molecule.

The oxygen atoms form an oxygen molecule. The chlorine atom is free to repeat the depletion process.

Figure 8 One chlorine atom can destroy nearly 100,000 molecules of ozone. If too many ozone molecules are destroyed, harmful radiation from the Sun could reach Earth.

Ozone Depletion About 20 km above Earth is the **ozone layer.** Ozone is a molecule made of three oxygen atoms, just like the ozone in smog. However, unlike smog, the ozone that exists at high altitudes helps Earth's organisms by absorbing some of the Sun's harmful rays.

In the mid-1980s, a severe depletion of ozone appeared over Antarctica. After researching, scientists discovered that CFCs are able to destroy ozone molecules, as shown in **Figure 8.**

In 1987, governments around the world agreed to restrict the use of CFCs gradually. By 1996, all industrialized nations halted production. Measurements taken in the upper atmosphere in 1996 show that the level of CFCs is beginning to decrease. However, scientists don't expect the ozone layer to recover until the middle of the twenty-first century.

section 1 review

Summary

What causes air pollution?

● Vehicles and industries cause air pollution.

Smog

● Smog forms when compounds react in sunlight.

Acid Rain

● Acid rain forms when compounds from burning fuel combine with moisture in the atmosphere.

Particulate Pollution

● Particulate matter includes suspended particles.

Toxic Pollutants and Carbon Monoxide

● Toxic pollutants might cause health problems.

Chlorofluorocarbons

● CFCs can damage Earth's ozone layer.

Self Check

1. **Compare and contrast** primary and secondary air pollutants.

2. **Explain** what causes photochemical smog.

3. **List** sources of toxic air pollution and carbon monoxide.

4. **Think Critically** Why would Denver, Colorado, have smoggier air on some days than Portland, Oregon?

Applying Math

5. **Solve One-Step Equations** If a person breathes air each day that contains 0.000 03 g of particulate matter, how many grams of particulate matter would a person breathe in one year?

 Science Online booki.msscience.com/self_check_quiz

PARTICULATE POLLUTION

You know that strong winds can pick up and carry small particles such as silt from soil. However, particles also can be carried in air from a variety of sources on most any day. In this activity, you will examine particulates in the air.

◗ Real-World Question

How much particulate matter is in the air? How do factors such as wind speed and location affect the amount of particulates?

Goals
■ **Observe** particulates in the air.
■ **Relate** particulate abundance to weather conditions and environment.

Materials
vacuum cleaner   compass
paper filters (5)   anemometer
*coffee filter   *local weather data source
stereo microscope   paper confetti
*hand lens   heavy-duty rubber bands (5)
*alternative materials

Safety Precautions 🖐️ 🥽 🗒️

WARNING: *Do not use the vacuum cleaner in the rain or on a wet surface. Keep the cord away from moisture.*

◗ Procedure

1. Find an outdoor location in your schoolyard where electrical power is available.
2. Each day for five days, determine the wind direction and speed at your location. Wind direction can be determined by dropping confetti and using a compass to determine the direction from which the wind is blowing. Wind speed can be determined using an

Particulate Data

Day	Weather Conditions	Particle Types and Number
1		
2	Do not write in this book.	
3		
4		
5		

anemometer. Alternately, the information can be obtained from a frequently updated source of weather information. Also, record other weather conditions such as snow cover and whether it rained recently.

3. Wrap a large, paper filter around the intake hose of a vacuum cleaner. Fasten the filter to the hose by tightly wrapping it with a heavy-duty rubber band.

4. Turn the vacuum cleaner on and let it draw air at a height of about 1 m for 20 min. each day. You should use a different filter on each different day of your experiment.

5. Remove the filter and take it indoors. Draw a circle around the area where particles would have collected.

6. Examine the filter under a stereomicroscope. Count and describe any particles you observe.

◗ Conclude and Apply

1. Describe the different types of particles collected on the filters.
2. Examine how wind conditions and other weather factors affected the number and type of particles collected each day.
3. Infer possible sources for the particles.

Effects of Air Pollution

What You'll Learn

- **Explain** how air pollution affects human health.
- **Describe** how air pollution affects Earth's organisms.
- **List** several ways that air pollution can damage buildings and structures.

Why It's Important

Air pollution can harm organisms, human-built structures, and your health.

🔍 Review Vocabulary

respiratory system: organs that take in oxygen and eliminate carbon dioxide

New Vocabulary

- ultraviolet radiation
- cataract
- biomagnification

Figure 9 Health effects of air pollution depend on the concentration of pollutants and how long you are exposed to them.

Air Pollution and Your Health

In the United States, more than 133 million people live in areas where air quality is unhealthy at times because of high levels of at least one pollutant. The United Nations estimates that at least 1.3 billion people around the world live in areas with dangerously polluted air.

The effects on your health from air pollution are listed in **Figure 9.** Health effects depend on how long you are exposed to the pollutant and how much of the pollutant is in the air. For example, you might notice watery eyes and shortness of breath on a smoggy day. When the air clears, you can breathe normally. If you breathe smoggy air for your entire life, you might have difficulty breathing when you get older.

Young children and elderly people suffer the most effects of pollution. With the same amount of exposure to pollutants, young children get much bigger doses for their size than adults do. When a child is young, all of his or her organs, including the brain, still are developing. Air pollution can affect the development of growing organs. Elderly people are at risk because they have been exposed to pollutants for a long time.

Short-Term Effects	Long-Term Effects
Stinging, watery eyes	Brain damage
Scratchy, sore throat	Liver disease
Cough	Kidney disease
Pneumonia	Lung cancer
Headache	Heart disease

Figure 10 Particles can lodge deep in your lungs and can damage your air sacs—the deepest part of your lung.

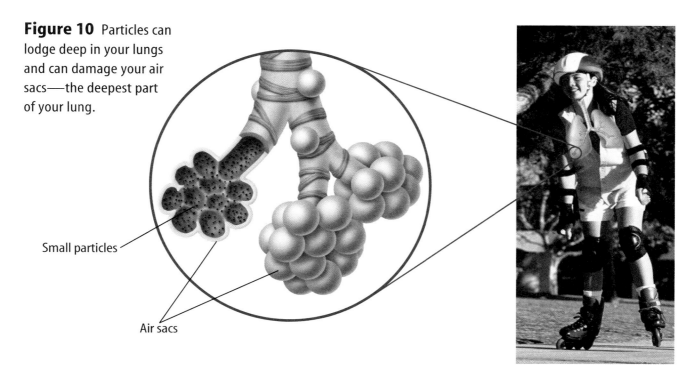

Small particles

Air sacs

Smog and Carbon Monoxide Compounds found in smog can cause your eyes to water and sting. Long-term exposure to smog can increase your risk for lung infections, reduce your ability to breathe normally, and might make asthma worse. You also can develop chest pains and a cough.

Carbon monoxide affects your blood's ability to carry oxygen. High concentrations of this gas might affect your vision, your ability to concentrate, and your coordination. Very high levels can cause death.

Effects of Particulates and Toxic Pollutants Do you sneeze when you shake out a dusty rug? When you sneeze, you force the dust out of your respiratory system. Smaller particles, however, can penetrate deep into your lungs and cause part of the lungs to become inflamed, as shown in **Figure 10.** Over time, small particulate matter might damage your lungs permanently, making breathing difficult and forcing your heart to work harder than it should.

Toxic substances in the air can damage many body systems. People exposed to toxic air pollutants can suffer from nerve damage, respiratory problems, and disorders of the reproductive system. They also can have an increased risk for cancer.

Exposure to a large amount of toxic air pollution over a short period of time can be deadly. On December 3, 1984, an accident at a pesticide factory in India released a cloud of toxic gas. More than 4,000 people died and 200,000 suffered permanent injuries such as blindness and heart disease.

INTEGRATE Health

Cigarette Smoking The tubes that carry air deep inside your lungs are lined with mucus-producing glands and small hair-like structures called cilia. The mucus traps particulate matter. The cilia move, constantly sweeping the mucus and the trapped particulate matter upward to your throat. Every time you swallow, you begin the process of clearing this mucus and particulate matter from your body. Cigarette smoking destroys the cilia over time. Infer what might happen to your lungs if you smoke cigarettes.

Figure 11 Malignant melanoma, a type of skin cancer that can be deadly, can be caused by exposure to the Sun's harmful rays.

Inhaling Acid When you inhale humid air from acid rain, acid can be deposited deep inside your lungs. Acid irritates the lung's sensitive tissues and reduces your ability to fight respiratory infections. Your lungs are responsible for moving oxygen into your blood. Damaged lungs cannot transfer oxygen to the blood easily, so the heart must work harder to pump oxygen to body cells. Over time, the heart can become stressed and weak.

Increased Ultraviolet Radiation Harmful rays from the Sun, called **ultraviolet radiation,** are blocked partially by the protective ozone layer. Each spring, an ozone hole forms over Antarctica. The hole is an area of the ozone layer that is thinning. The size of the hole changes from year to year because of temperature variations in the atmosphere.

In humans, increased ultraviolet radiation is linked to skin cancer. One type of skin cancer, malignant melanoma, shown in **Figure 11,** accounts for only four percent of skin cancer cases but causes about 79 percent of skin cancer deaths. About 54,000 people are diagnosed with malignant melanoma in the United States each year, and nearly 7,600 people die from it. The number of new melanomas diagnosed has more than doubled since 1973.

In addition to skin cancer, cataracts are more common in people who are exposed to high amounts of ultraviolet radiation. **Cataracts** are a form of eye damage that make the lens of the eye cloudy. Ultraviolet radiation also can affect the immune system, which helps you fight illness.

✔ **Reading Check** *What are some health effects of increased ultraviolet radiation?*

You can protect yourself against excess ultraviolet radiation by avoiding outdoor activities during the middle of the day and by wearing long sleeves, a hat with a wide brim, sunglasses, and by using protective sunscreens.

Effects on Earth's Organisms

Animals are exposed to air pollutants when they inhale gases and small particles. Because air pollutants fall to Earth in rain or snow, animals also are exposed when they ingest pollutants in their food and water. Soft-bodied animals such as earthworms, or animals with thin, moist skin, such as amphibians, can absorb air pollutants directly through their skin.

Just like humans, young animals are not able to tolerate the same amount of pollution as adult animals can. Whether or not an animal will be affected by a pollutant depends on the kind of pollutant, the length of time the animal is exposed to the pollutant, and the amount of pollutant taken into the animal's body.

Concentrating Pollutants The concentration of a pollutant in the air might not be high enough to cause a problem. Some pollutants, however, stay in animal tissues instead of being excreted from their bodies as waste. When these animals are eaten by other animals, the pollutants are passed on to the predator. **Biomagnification** (BI oh mag nuh fuh KAY shun) is the process in which pollutant levels increase through the food chain, as shown in **Figure 12.** Some fish are not safe for humans to eat frequently because of biomagnification.

Figure 12 Pollutants from the air, such as the metal mercury, can end up in high concentrations in animals through biomagnification. Pollutants often are measured in parts per million, or ppm.

Figure 13 Acids in rain or fog can make trees weak and strip the nutrients they need from the soil.

Acidic Lakes and Streams Recall that lower pH means higher acidity. The pH of some streams, lakes, and rivers can decrease when acid rain falls. Many organisms require an environment with a narrow range of pH values.

In some streams and lakes in the United States and Canada, acid rain has eliminated certain fish species, such as brook trout. For example, hundreds of lakes in the Adirondack Mountains in New York are too acidic for the survival of fish. The Canadian government estimates that more than 14,000 lakes in eastern Canada are acidic. Acid rain is an even greater problem when snow melts. If a large amount of acidic snow falls in the winter and melts quickly in the spring, a sudden rush of acids flows into lakes and streams. Many fish and other organisms have been killed because of sudden pH changes.

Acid rain also can damage plants. At higher elevations, trees often are surrounded by fog. When the fog is acidic, trees suffer injury and are less able to resist pests and diseases. Some stands of evergreens in the Great Smoky Mountain National Park, as shown in **Figure 13,** have died from acidic exposure.

Acid Rain and Soils Acid rain can also affect soils. As acid rain moves through soil, it can strip away many of the nutrients that trees and other plants need to grow. Some regions of the United States, however, have naturally basic soils. In such regions, acid rain might not significantly affect vegetation. The higher pH of basic soils can help raise the pH of acid rain after it falls to the ground.

Smog The compounds in smog affect animals and plants. Smog affects the respiratory systems of animals, causing irritation to the lining of the lungs. When plants are exposed to smog over a long period of time, the pollutants break down the waxy coating on their leaves. This results in water loss through the leaves and increases the effects of diseases, pests, drought, and frost. Scientists estimate that smog formed from vehicle exhaust damages millions of dollars worth of crops in California each year.

 **Reading Check** *What effects does smog have on plants and animals?*

The Ozone Layer As the ozone layer thins, Earth's organisms are exposed to more ultraviolet radiation. Small organisms called phytoplankton (FI tuh PLANG tun) live in Earth's freshwater and oceans. They make food using carbon dioxide and water in the presence of sunlight. These organisms also are the basis of the food chain shown in **Figure 14.** Research shows that ultraviolet radiation can reduce the ability of phytoplankton to make food, decreasing their numbers. Ultraviolet radiation also might damage young crabs, shrimp, and some fish. In some animal species, growth is slowed and the ability to fight diseases is reduced.

Ultraviolet radiation might affect many agricultural crops such as rice by decreasing the plant's ability to fight diseases and pests. Even small increases in ultraviolet radiation might reduce the amount of rice grown per square kilometer. Rice is the main food source for more than half the world's population. With world population increasing, a crisis might occur if rice and other crop production is affected by ultraviolet radiation.

Reading Check *How might increasing ultraviolet radiation affect world rice production?*

Figure 14 An increase in ultraviolet radiation could decrease the number of fish in antarctic waters.

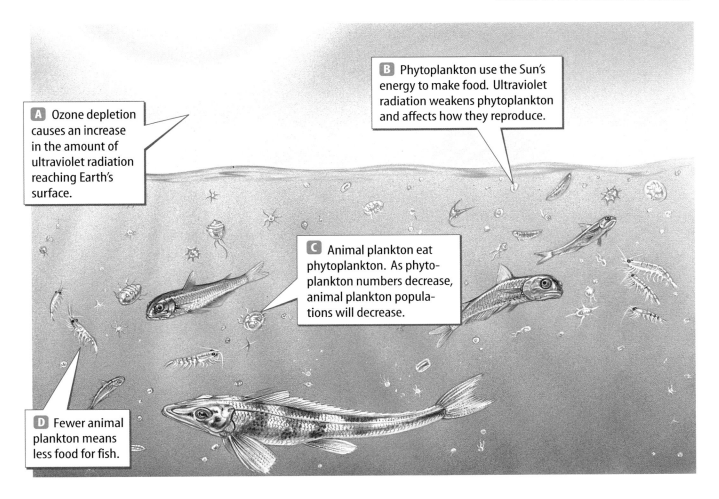

A Ozone depletion causes an increase in the amount of ultraviolet radiation reaching Earth's surface.

B Phytoplankton use the Sun's energy to make food. Ultraviolet radiation weakens phytoplankton and affects how they reproduce.

C Animal plankton eat phytoplankton. As phytoplankton numbers decrease, animal plankton populations will decrease.

D Fewer animal plankton means less food for fish.

Figure 15 The pyramids in Egypt have withstood the Sun, wind, and sandstorms for more than 4,000 years. However, air pollution within the last 50 years has led to increased decay of these magnificent structures.

Damage to Materials and Structures

Air pollution not only affects your health and the health of other organisms, it also damages many materials. For example, acid rain is known to corrode metals and deteriorate stone and paint. To reduce the damage on automobiles, some manufacturers use a very expensive, acid-resistant paint. Smoke and soot coat buildings, paintings, and sculptures, requiring expensive cleaning. In cities all over the world, works of art, ornate buildings and statues, and structures like the pyramids of Egypt, shown in **Figure 15,** suffer from the effects of air pollution.

section 2 review

Summary

Air Pollution and Your Health

- Smog can cause breathing problems.
- Particulate pollution can damage your lungs.
- Ultraviolet radiation is linked to skin cancer.

Effects on Earth's Organisms

- Biomagnification increases pollutants through the food chain.
- Many lakes have no fish due to acid rain.
- Smog can damage leaves, causing plants to lose water.

Damage to Materials and Structures

- Structures require expensive cleaning because of soot and smoke.
- Acid rain corrodes metals.

Self Check

1. **List** three ways that animals are exposed to air pollutants.
2. **Summarize** the effects of air pollution on human health.
3. **Infer** why young children are more affected by air pollutants than young adults are.
4. **Think Critically** What might happen to carbon dioxide levels in Earth's atmosphere if ultraviolet radiation increases?

Applying Skills

5. **Use graphics software** to create a pamphlet that describes the effects that acid rain might have on humans and other organisms.

 Science Online booki.msscience.com/self_check_quiz

Solutions to Air Pollution

Clean Air Laws

Between 1900 and 1970, motor vehicle use and industrial manufacturing grew rapidly in the United States. Air in some parts of the country, especially in cities, became more polluted. Nitrogen oxides, which help form smog and acid rain, increased nearly 1,000 percent between 1900 and 1970.

Scientists and government officials recognized that air quality must be protected. Beginning in 1955, the U.S. Congress passed a series of laws to help protect the air you breathe. A summary of these laws is listed in **Table 1.** The U.S. Environmental Protection Agency has the responsibility of gathering and analyzing air pollution data from across the country and working to keep the country's air clean.

The Clean Air Act is a federal law that regulates air pollution over the entire country. Each state is responsible for making sure that the goals of the law are met. State agencies limit what power plants and industries can release into the air. Companies that exceed air pollution limits might have to pay a fine. Automobile exhaust is monitored in areas with poor air quality.

as you read

What You'll Learn

■ **Describe** air pollution laws in the United States.
■ **Identify** things you can do to reduce air pollution.

Why It's Important

Controlling the sources of air pollution will help keep your air clean.

Review Vocabulary
coal: sedimentary rock formed from decayed plant material; the world's most abundant fossil fuel

New Vocabulary
● ambient air
● air quality standard
● emission

Table 1 Summary of Clean Air Regulations in the United States	
Name of Law	**What It Does**
Air Pollution Control Act of 1955	It granted $5 million annually for air pollution research. Although it did little to prevent air pollution, the law made the public aware of pollution problems.
Clean Air Act of 1963	This act granted $95 million per year to state and local governments for research and to create air pollution control programs. It also encouraged the use of technology to reduce air pollution from cars and electric power plants.
Clean Air Act of 1970	It set standards for specific pollutants in the air and placed strict limits on car exhaust and pollutants from new industries.
Clean Air Act of 1990	This act placed strict limits on car emissions and encouraged the use of cleaner-burning gasoline. It also forced companies to reduce toxic emissions.

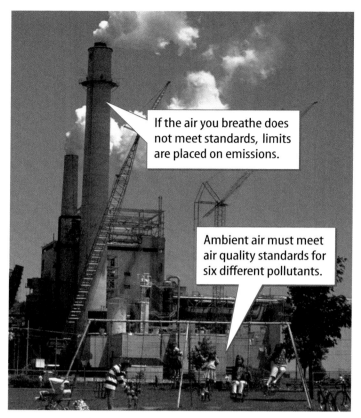

If the air you breathe does not meet standards, limits are placed on emissions.

Ambient air must meet air quality standards for six different pollutants.

Ambient Air You know that air pollutants released in one part of the country can affect the air somewhere else. Natural events, such as temperature inversions, can concentrate pollutants in an area. The surrounding air you breathe is called **ambient** (AM bee unt) **air.** Air pollution laws are written to help keep ambient air clean, no matter what the source of pollution is. Across the United States, scientists sample and test ambient air for particulate matter, carbon monoxide, sulfur dioxide, nitrogen dioxide, lead, and ozone. These pollutants cannot exceed a certain level, called an **air quality standard.** Areas that already have fairly clean air, such as national parks, have stricter air quality standards than cities do. As **Figure 16** illustrates, if an area has pollution levels above ambient air quality standards, more controls can be placed on specific sources of air pollution.

Figure 16 The Clean Air Act of 1990 set standards for ambient air. These limits are met by controlling emissions from different sources.

Topic: Air Quality Standards

Visit booki.msscience.com for Web links to information about air quality standards.

Activity Research the air quality guide for ozone. Record the air quality index of a city near you for 10 days.

Controlling the Source Pollutants released into the air from a particular source are called **emissions** (ee MIH shunz). Emissions are measured at industry smokestacks and automobile tailpipes. If ambient air quality standards are not met, emissions must be reduced.

Emissions can be controlled in two ways—by using devices that capture pollutants already created and by limiting the amount of pollutants produced in the first place. For example, auto exhaust used to contain many more pollutants than it does today. Since 1975, each new car sold in the United States has been equipped with a catalytic (ka tuh LIH tihk) converter, a device that changes harmful gases in car exhaust to less harmful ones. A catalytic converter and other emission control devices are shown in **Figure 17.**

 **Reading Check** *What two methods can be used to control emissions?*

Changing the way gasoline is produced has helped control the amount of pollutants in gasoline even before it is burned. Compounds such as alcohol can be added to gasoline to reduce tailpipe emissions. Since the 1990 Clean Air Act was enacted, only clean-burning gasoline can be sold in the smoggiest areas of the country.

NATIONAL GEOGRAPHIC

VISUALIZING EMISSIONS CONTROL

Figure 17

In the past few decades new technologies have reduced air pollution by trapping pollutants at their sources. Devices such as smokestack scrubbers, electrostatic precipitators, and catalytic converters shown here use different methods to remove pollutants from exhaust gases.

◀ **SMOKESTACK SCRUBBER** Burning some types of coal to generate electricity produces large quantities of sulfur dioxide—a pollutant that can cause acid rain. The smokestacks of many coal-burning plants are equipped with antipollution devices called scrubbers.

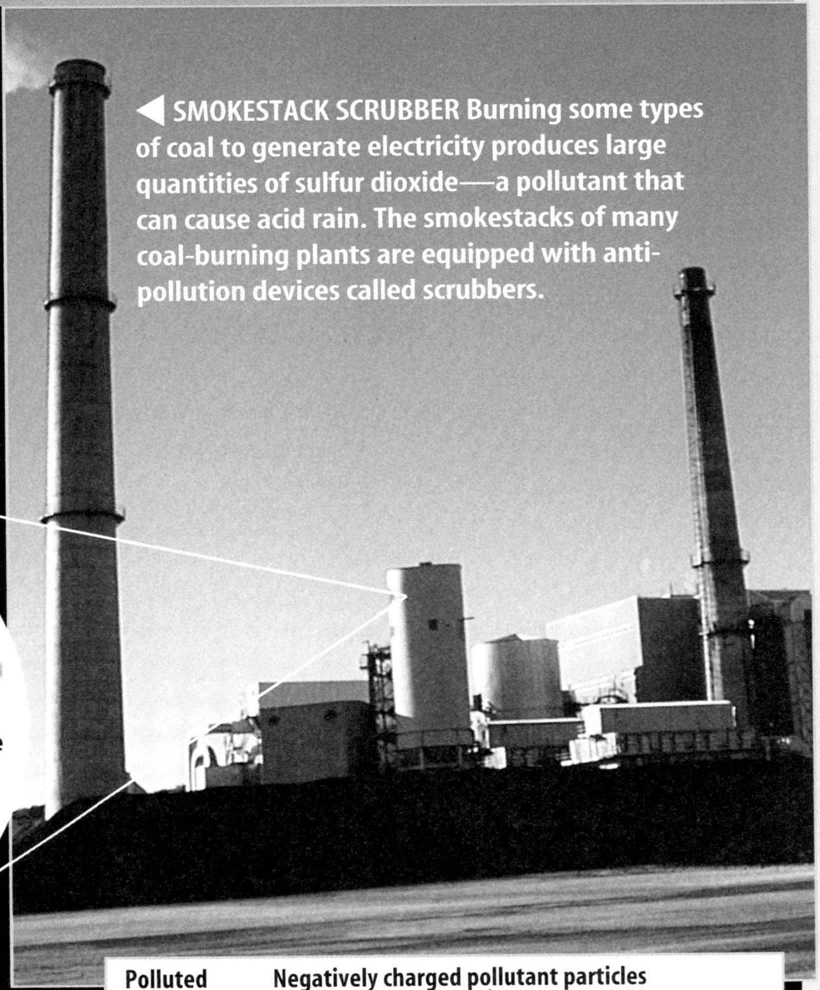

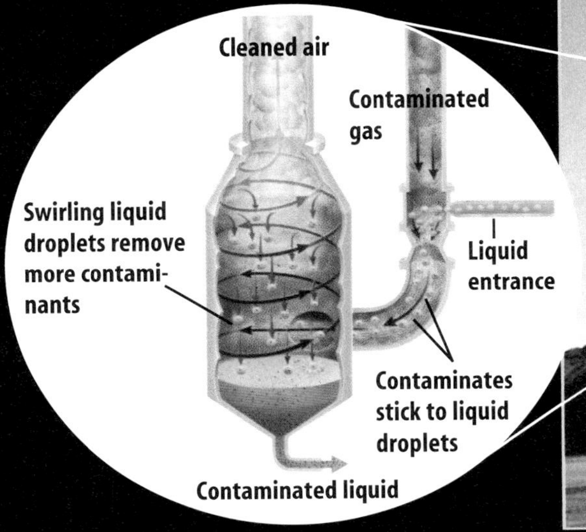

Cleaned air
Contaminated gas
Swirling liquid droplets remove more contaminants
Liquid entrance
Contaminates stick to liquid droplets
Contaminated liquid

▶ **ELECTROSTATIC PRECIPITATOR** As smoke enters an electrostatic precipitator, plates that line the interior of the device give polluting particles a negative charge. Positively charged plates then attract the particles, "cleaning" the smoke. An electrostatic precipitator removes up to 99 percent of particulate matter from industrial emissions.

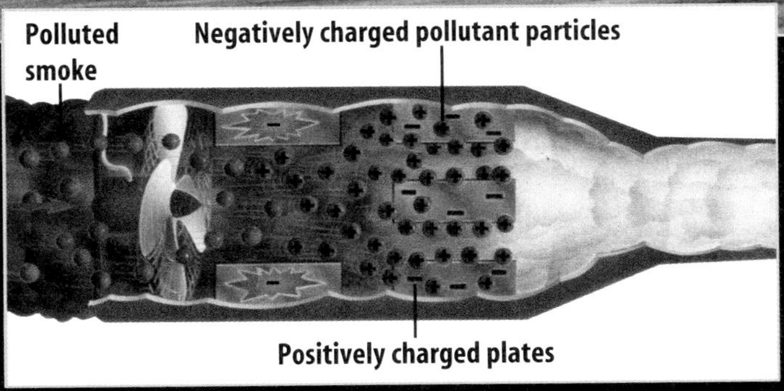

Polluted smoke
Negatively charged pollutant particles
Positively charged plates

◀ **CATALYTIC CONVERTER** Automobile exhaust gases pass over small beads coated with metals inside a catalytic converter. The metals cause chemical reactions that change most of the harmful gases into carbon dioxide and water.

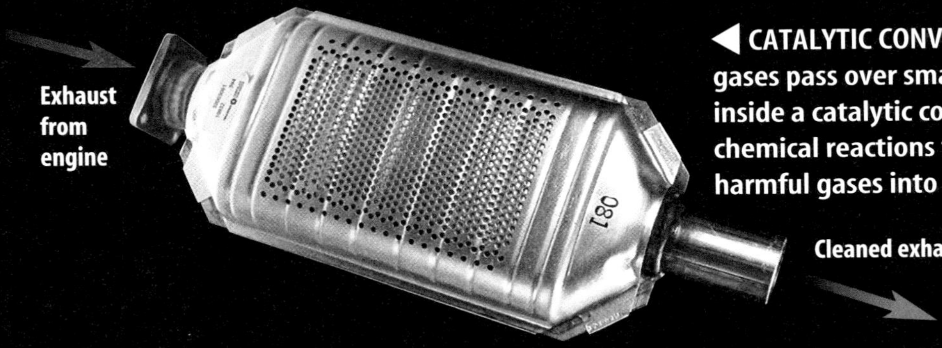

Exhaust from engine
Cleaned exhaust exits tailpipe

Figure 18 Cars emit 0.6 g of nitrogen oxides per kilometer. Light trucks, minivans, and sport-utility vehicles can emit 1.1 g to 1.7 g per kilometer, depending on their size.

Calculate *How many grams of nitrogen oxides would not be emitted if you rode your bike for 5 km instead of riding in a car?*

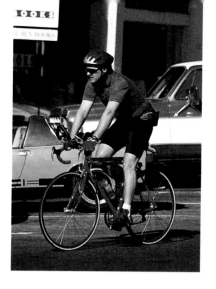

You Can Help

Laws and new technologies will help reduce air pollution, but you can be a part of the solution, too. When you reduce the amount of electricity you use, less fuel is burned at a power plant, and less pollution is released. Turn off lights and all appliances when you aren't using them. Turn down the thermostat in the winter and wear more layers of clothing. Open windows in the summer instead of using air conditioning. Using public transportation, riding a bike, as shown in **Figure 18,** or car pooling will help keep the air clean.

✔ **Reading Check** *How can you help reduce air pollution?*

Applying Math   Solve a One-Step Equation

BURNING COAL Sulfur dioxide (SO_2) forms when coal or oil is burned. It is considered to be a major air pollutant. Burning a certain type of coal produces about 0.01 kg of SO_2 per kilogram of coal. If a power plant burns 3 million kg of coal annually, how much SO_2 would be released?

Solution

1 *This is what you know:*
- production rate = 0.01 kg of SO_2 /kg coal
- annual coal use = 3,000,000 kg

2 *This is what you need to find:*

annual emissions = kilograms of SO_2 produced each year

3 *This is the equation you need to use:*
- annual emissions = (annual use) × (production rate)
- annual emissions = (3,000,000 kg coal) × (0.01 kg SO_2 /kg coal) = 30,000 kg SO_2

4 *Check your answer:*

Divide your answer by the annual coal use. You should get the production rate for SO_2

Practice Problems

1. If a power plant burned 500,000 kg of coal annually, how much SO_2 would be produced?

2. If coal contained 0.02 kg of SO_2 per kg of coal, how much SO_2 would be produced if 250,000 kg of coal was burned?

Science Online
For more practice, visit booki.msscience.com/ math_practice

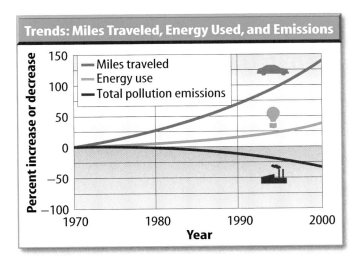

Trends: Miles Traveled, Energy Used, and Emissions

- Miles traveled
- Energy use
- Total pollution emissions

Figure 19 The total amount of air pollutants released to the air above the United States has decreased since the passage of strict air pollution laws.

Improving Air Quality

As **Figure 19** shows, air quality in the United States has improved since 1990, even though energy use increased and people are driving more. Strict controls on sources of pollution have greatly increased the quality of the air you breathe. Even so, although the national trends for most air pollutants are decreasing, some others, such as nitrogen dioxides, continue to rise. Smog levels are increasing in many rural areas, and haze is a problem in some national parks. The United States is home to more than 2,500 bodies of water whose fish are unsafe to eat because of biomagnification of toxins. As the United States population continues to increase, conservation and new technology can help reduce air pollution.

section 3 review

Summary

Clean Air Laws

- Laws passed in the United States since 1955 have helped control air pollution.
- Air pollution laws are written to help keep ambient air clean.
- When ambient air becomes too polluted, controls are placed on emissions.

You Can Help

- Conserving electricity will help reduce air pollution.
- Riding your bike can help keep the air clean.

Self Check

1. **List** six pollutants sampled in ambient air.
2. **Explain** what the 1970 Clean Air Act regulated.
3. **Infer** how air pollution is reduced when energy is conserved.
4. **Think Critically** Why are air pollution standards written for ambient air instead of emissions?

Applying Math

5. **Communicate** Write a letter to the editor of a newspaper explaining how everyone can help reduce air pollution.

LAB

Use the Internet

Air Pollution Where You Live

Goals

- **Identify** the air quality index value and weather conditions for a city near you for a specified time period.
- **Evaluate** trends in weather patterns and air quality.
- **Draw conclusions** about how weather patterns affect air quality.

Data Source

Visit booki.msscience.com/ internet_lab to get more information about air quality and for data collected by other students.

◉ Real-World Question

The quality of the air you breathe can affect your health and the health of other organisms near your home. Clean air laws passed in 1970 and 1990 have helped improve air quality in many regions of the United States. However, landforms and weather can affect air quality.

◉ Form a Hypothesis

The air quality index tells you how clean the air is and whether it will affect your health. Some areas of the United States experience more air quality problems than others. Form a hypothesis about how weather conditions affect the air quality index.

◉ Make a Plan

1. **Research** information about the air quality index. Data on the types of air pollutants collected by federal or state sources can be helpful.
2. **Research** specific weather data that can affect air pollution levels.
3. **Investigate** weather patterns that can contribute to an increase in the air quality index for a city or region.
4. Look for other types of information that provide additional clues about your community's air quality. For example, do gasoline stations in your area have special equipment to prevent fumes from escaping while fueling your car?

Air Quality	Air Quality Index	Protect Your Health
Good	0–50	No health impacts occur.
Moderate	51–100	People with breathing problems should limit outdoor exercise.
Unhealthy for certain people	101–150	Everyone, especially children and elderly, should not exercise outside for long periods of time.
Unhealthy	151–200	People with breathing problems should avoid outdoor activities.
Very Unhealthy	201–300	Health alert: everyone may experience more serious health effects.
Hazardous	>300	Health warnings of emergency conditions.

▶ Follow Your Plan

1. Make sure your teacher approves your plan before you start.
2. Create a table for your data.

▶ Analyze Your Data

1. **List** the weather data and air quality index values you collected during the time period.
2. **Compare and contrast** weather data and the air quality index values.
3. **Graph** your data. You may want to show how temperature and the air quality index are related, or you may want to illustrate how values changed during the time period that you collected data.

▶ Conclude and Apply

1. **Evaluate** Overall, how would you assess the quality of your community's air?
2. **Determine** Is there a relationship between weather data and air quality data?
3. **Discuss** patterns in the air quality index over several days or weeks. For example, air quality may worsen from Monday to Friday. Explain why this might be happening.

𝒞ommunicating Your Data

Find this lab using the link below. Post your data in the table provided. Compare your data to those of other students and plot the data on a graph. Present your findings to your class.

Science Online
booki.msscience.com/internet_lab

Radon

The Invisible Threat

How do you protect yourself from something you can't see, smell, or taste? Years ago, the famous scientist Marie Curie frequently handled radioactive materials. Curie was investigating their properties, never realizing that the invisible rays emitting from the substances were slowly poisoning her. In 1934, she died of leukemia, which was most likely brought on by exposure to radiation.

Scientists eventually realized the danger of radioactive materials. Many workers in uranium mines developed lung cancer as a result of being exposed to radioactive substances. Scientists also learned that as uranium begins to break down, it changes into different elements, such as thorium, protactinium (pro tak TIH nee um), and radium . . . all still radioactive.

In 1900, a German scientist, Friedrich Ernst Dorn, discovered that radium emitted a radioactive gas called radon. Still, most people did not think radon gas was much to worry about.

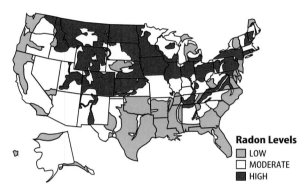

Radon Levels
- LOW
- MODERATE
- HIGH

This map shows radon risk in the United States.

All that changed in 1984. An engineer at a nuclear power plant in Pennsylvania set off the radiation detectors at the plant one morning. Officials found no contamination that could have caused this to happen. So, where did the engineer get his radioactive contamination? The answer was completely unexpected—the engineer's home! The house had radiation levels 700 times higher than is considered to be safe for humans. Further study found that the house was built on rock that contained uranium and radon gas.

Fortunately, there are tests that people living in high-risk radon areas can conduct to detect levels of radon gas. And, if found, there are methods for removing radon gas from these buildings safely.

Marie Curie was exposed to radioactive materials.

Use a map Locate the area where you live on the map above. Are high amounts of radon gas found in your state? According to the map, which states have low radon levels?

Science online

For more information, visit booki.msscience.com/time

Reviewing Main Ideas

Section 1 **Types and Causes of Air Pollution**

1. Human activities and nature can cause air pollution.

2. Acid rain forms when compounds combine with moisture in the atmosphere.

3. Mountains, weather, and temperature inversions can add to smog development.

4. Toxic air pollutants come from vehicles, businesses, factories, power plants and volcanoes.

Section 2 **Effects of Air Pollution**

1. Depending on the exposure, smog, acid rain, and particulate pollution can cause minor discomfort or lead to long-term health problems.

2. Animals are exposed to air pollutants when they breathe polluted air. Some animals also can absorb pollutants through their skin.

3. The ozone layer protects Earth from ultraviolet radiation.

4. Pollutants can increase in concentration as they biomagnify through food chains.

Section 3 **Solutions to Air Pollution**

1. Air pollution laws passed since 1955 have reduced air pollution levels.

2. Catalytic converters on cars and smokestacks scrubbers help reduce air pollution.

Visualizing Main Ideas

Copy and complete the following chart on the health effects of air pollution.

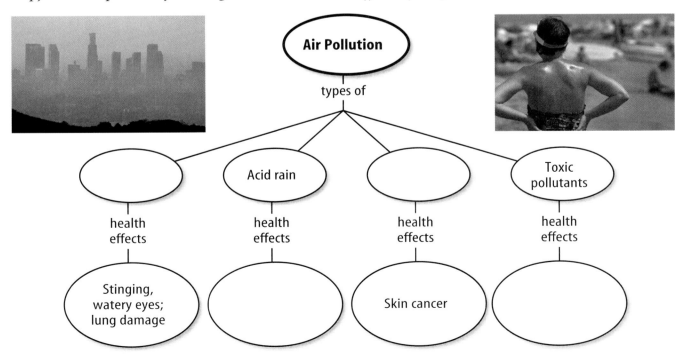

Air Pollution

types of

Acid rain

Toxic pollutants

health effects — Stinging, watery eyes; lung damage

health effects

health effects — Skin cancer

health effects

Using Vocabulary

acid rain p. 98	particulate matter p. 100
air quality standard p. 112	photochemical
ambient air p. 112	smog p. 97
biomagnification p. 107	primary pollutant p. 96
cataract p. 106	secondary pollutant p. 96
emission p. 112	toxic air pollutant p. 101
ozone layer p. 102	ultraviolet radiation p. 106

Use what you know about the listed terms to answer the following questions.

1. How do pollutants form photochemical smog? Give an example of a secondary pollutant in smog.

2. Why are emissions controlled when ambient air doesn't meet an air quality standard?

3. Why didn't the ozone layer immediately recover despite the severe restriction of chlorofluorocarbon use in the mid-1990s?

4. Explain how sulfur dioxide emissions in the Midwest contribute to acid rain in the Northeast.

5. How can biomagnification of toxic air pollutants occur in the environment?

Checking Concepts

Choose the word or phrase that best answers the question.

6. What is the brown haze that forms over some cities called?
   A) CFCs
   B) carbon monoxide
   C) acid rain
   D) smog

7. What is the pH of acid rain?
   A) greater than 7.6   C) lower than 5.6
   B) 5.6   D) 5.6 to 7.6

8. What does the ozone layer absorb?
   A) metals   C) acid rain
   B) UV radiation   D) particulates

9. What type of pollutant is about one-seventh the diameter of a human hair?
   A) coarse particulate matter
   B) fine particulate matter
   C) acid rain
   D) carbon monoxide

10. Which term is used to describe increasing pollutant levels through the food chain?
    A) ambient   C) biomagnification
    B) emission   D) acidity

11. Most carbon monoxide pollution comes from which source?
    A) power plants
    B) cleaning products
    C) car exhaust
    D) industry

12. Which emission control device was added to automobiles in the mid-1970s?
    A) catalytic converter
    B) smokestack scrubber
    C) alcohol
    D) electrostatic separator

Use the figure below to answer question 13.

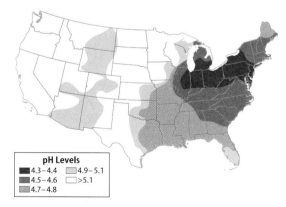

pH Levels
- 4.3–4.4
- 4.5–4.6
- 4.7–4.8
- 4.9–5.1
- >5.1

13. Which state has precipitation that is most acidic?
    A) California   C) Virginia
    B) Pennsylvania   D) Louisiana

Thinking Critically

14. **Describe** how traffic jams can increase air pollution.

15. **Explain** why air pollution affects the health of older people more than it affects most middle-aged people.

16. **Infer** why there are air quality standards for ambient air.

17. **List** three things you can do to help reduce air pollution.

18. **Recognize Cause and Effect** Why is the concentration of some pollutants greater in birds and mammals than in the contaminated organisms they eat?

19. **Recognize Cause and Effect** It has been sunny and hot with little wind for more than a week in your city. The newspaper reports that smog levels are unhealthy. Explain the connection.

20. **Compare and contrast** the effects of acid rain and ultraviolet radiation on organisms.

21. **List** some ways you can protect yourself against exposure to ultraviolet radiation.

22. **Explain** why air in some parts of the United States became more polluted from 1900 to 1970.

23. **Recognize Cause and Effect** After a quick spring thaw of heavy snow, many fish are found dead in a lake. Explain why this might happen.

24. **Describe** three kinds of air pollution caused by burning fuel in vehicles.

25. **Explain** how chlorofluorocarbon atoms in the atmosphere can increase ultraviolet radiation at Earth's surface.

26. **Explain** how electrostatic precipitators remove particulate matter from smoke.

Performance Activities

27. **Research Information** Research the pollutants found in cigarette smoke. Make a poster of how these substances harm your health.

28. **Organize Activities** To conserve energy in your school, place signs near light switches that remind people to turn off lights when they leave the room. Make signs encouraging students to carpool to school events.

29. **Design and perform an experiment** to test the effects of acid rain on plants. Remember to test one variable at a time.

Applying Math

30. **Convert Units** Most air pollutants are measured in micrograms. There are 1,000 micrograms in 1 milligram and 1,000 milligrams in 1 gram. How many micrograms are in 1 gram?

Use the figure below to answer questions 31 and 32.

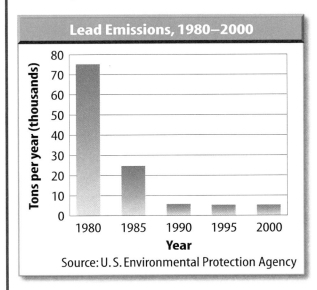

Lead Emissions, 1980–2000

Source: U.S. Environmental Protection Agency

31. **Decrease over Five Years** Estimate the percent decrease in lead emissions from 1980 to 1985.

32. **Decrease over Twenty Years** Estimate the total decrease in lead emissions from 1980 to 2000.

Part 1 Multiple Choice

Record your answers on the answer sheet provided by your teacher or on a sheet of paper.

1. Which are harmful substances released directly into the air?
   A. nitrogen
   B. oxygen
   C. primary pollutants
   D. secondary pollutants

2. Which are harmful substances that form in the atmosphere?
   A. nitrogen
   B. oxygen
   C. primary pollutants
   D. secondary pollutants

Use the graph below to answer questions 3–4.

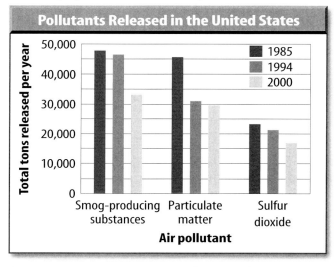

3. How many more tons of particulate matter were released in the United States in 1985 than in 2000?
   A. 20,000   C. 10,000
   B. 30,000   D. 15,000

4. About how many total tons of pollutants listed in the figure were released in 2000?
   A. 80,000   C. 40,000
   B. 120,000   D. 60,000

5. What is the reddish-brown gas that contributes to the colored haze of smog?
   A. carbon dioxide
   B. nitrogen dioxide
   C. oxygen
   D. ozone

Use the graph below to answer questions 6–7.

Sources of Human-Caused Toxic Air Pollution

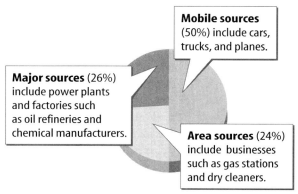

Mobile sources (50%) include cars, trucks, and planes.

Major sources (26%) include power plants and factories such as oil refineries and chemical manufacturers.

Area sources (24%) include businesses such as gas stations and dry cleaners.

6. Which is considered a major source of toxic air pollution?
   A. gas stations
   B. cars, truck and planes
   C. dry cleaners
   D. factories and chemical manufacturers

7. Which source emits the highest percentage of toxic air pollution?
   A. cars, truck and planes
   B. dry cleaners
   C. factories and chemical manufacturers
   D. gas stations

8. Which federal law regulates air pollution over the entire country?
   A. Air Pollution Control Act
   B. Clean Air Act
   C. Environmental Protection Act
   D. Safe Air Act

Part 2 | Short Response/Grid In

Record your answers on the answer sheet provided by your teacher or on a sheet of paper.

9. List some short-term health problems caused by polluted air. Who is most at risk for developing long-term effects?

10. List two sources from nature that can contribute to air pollution.

Use the figure below to answer questions 11–12.

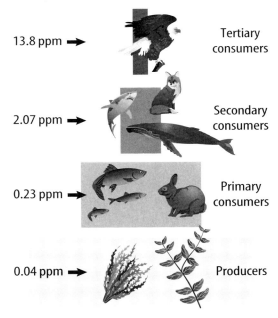

13.8 ppm ➡ Tertiary consumers

2.07 ppm ➡ Secondary consumers

0.23 ppm ➡ Primary consumers

0.04 ppm ➡ Producers

11. Which process is illustrated in the figure?

12. Explain how pollutant levels in organisms increase through the food chain.

13. Why is acid rain more of a concern as winter ends?

14. Why and for which pollutants is ambient air tested across the United States?

15. What would be the effect of increasing ultraviolet radiation on corn production?

16. To what is the air quality standard applied? How is this done?

Part 3 | Open Ended

Record your answers on a sheet of paper.

17. What must be done if local pollution levels are above ambient air quality standards?

18. Under which conditions might acid rain have less of an effect on the environment?

19. How do chlorofluorocarbons destroy ozone?

Use the figure below to answer questions 20–21.

Comparison of Growth Areas and Emission Trends

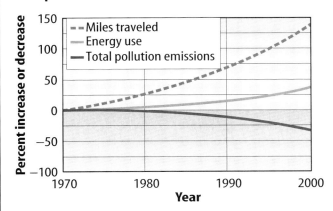

20. The figure above shows the percent of air pollution over the US from 1970 to 2000. What has led to the decrease in total pollution emissions?

21. What evidence from the figure supports the claim that use of catalytic converters has reduced automobile pollutant emissions?

22. How do landforms affect smog?

23. Limestone is a rock that is basic in nature and is known for its ability to neutralize the effects of acid rain. How could limestone be implemented into a pollution reduction program?

24. How could acid rain be responsible for a reduction in the ability to fight infections or for weakening of the heart muscle in humans?

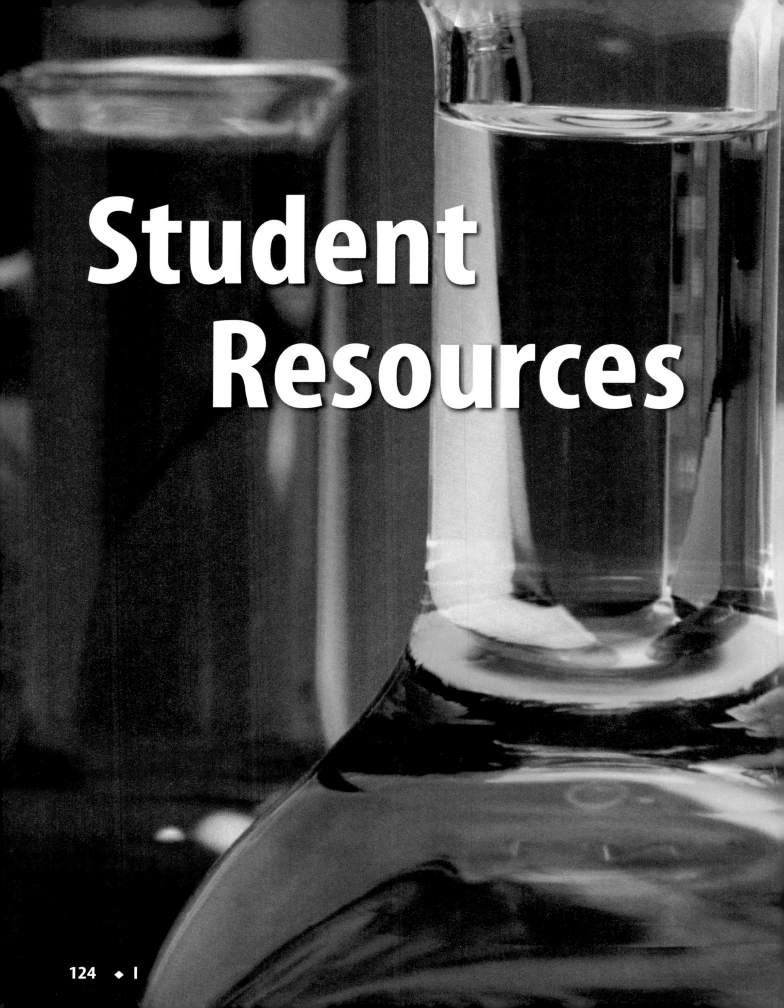

Student Resources

CONTENTS

Scientific Methods

Scientists use an orderly approach called the scientific method to solve problems. This includes organizing and recording data so others can understand them. Scientists use many variations in this method when they solve problems.

Identify a Question

The first step in a scientific investigation or experiment is to identify a question to be answered or a problem to be solved. For example, you might ask which gasoline is the most efficient.

Gather and Organize Information

After you have identified your question, begin gathering and organizing information. There are many ways to gather information, such as researching in a library, interviewing those knowledgeable about the subject, testing and working in the laboratory and field. Fieldwork is investigations and observations done outside of a laboratory.

Researching Information Before moving in a new direction, it is important to gather the information that already is known about the subject. Start by asking yourself questions to determine exactly what you need to know. Then you will look for the information in various reference sources, like the student is doing in **Figure 1.** Some sources may include textbooks, encyclopedias, government documents, professional journals, science magazines, and the Internet. Always list the sources of your information.

Figure 1 The Internet can be a valuable research tool.

Evaluate Sources of Information Not all sources of information are reliable. You should evaluate all of your sources of information, and use only those you know to be dependable. For example, if you are researching ways to make homes more energy efficient, a site written by the U.S. Department of Energy would be more reliable than a site written by a company that is trying to sell a new type of weatherproofing material. Also, remember that research always is changing. Consult the most current resources available to you. For example, a 1985 resource about saving energy would not reflect the most recent findings.

Sometimes scientists use data that they did not collect themselves, or conclusions drawn by other researchers. This data must be evaluated carefully. Ask questions about how the data were obtained, if the investigation was carried out properly, and if it has been duplicated exactly with the same results. Would you reach the same conclusion from the data? Only when you have confidence in the data can you believe it is true and feel comfortable using it.

Interpret Scientific Illustrations As you research a topic in science, you will see drawings, diagrams, and photographs to help you understand what you read. Some illustrations are included to help you understand an idea that you can't see easily by yourself, like the tiny particles in an atom in **Figure 2.** A drawing helps many people to remember details more easily and provides examples that clarify difficult concepts or give additional information about the topic you are studying. Most illustrations have labels or a caption to identify or to provide more information.

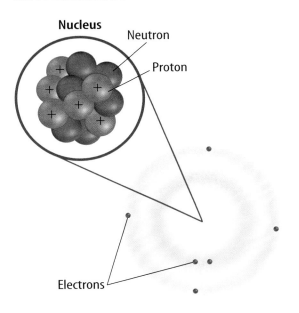

Figure 2 This drawing shows an atom of carbon with its six protons, six neutrons, and six electrons.

Concept Maps One way to organize data is to draw a diagram that shows relationships among ideas (or concepts). A concept map can help make the meanings of ideas and terms more clear, and help you understand and remember what you are studying. Concept maps are useful for breaking large concepts down into smaller parts, making learning easier.

Network Tree A type of concept map that not only shows a relationship, but how the concepts are related is a network tree, shown in **Figure 3.** In a network tree, the words are written in the ovals, while the description of the type of relationship is written across the connecting lines.

When constructing a network tree, write down the topic and all major topics on separate pieces of paper or notecards. Then arrange them in order from general to specific. Branch the related concepts from the major concept and describe the relationship on the connecting line. Continue to more specific concepts until finished.

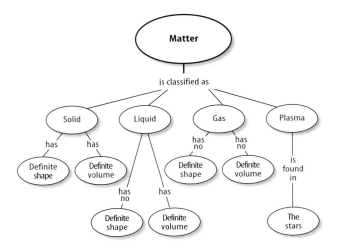

Figure 3 A network tree shows how concepts or objects are related.

Events Chain Another type of concept map is an events chain. Sometimes called a flow chart, it models the order or sequence of items. An events chain can be used to describe a sequence of events, the steps in a procedure, or the stages of a process.

When making an events chain, first find the one event that starts the chain. This event is called the initiating event. Then, find the next event and continue until the outcome is reached, as shown in **Figure 4.**

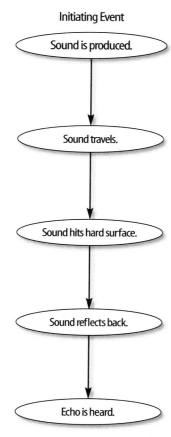

Initiating Event

Sound is produced.

Sound travels.

Sound hits hard surface.

Sound reflects back.

Echo is heard.

Figure 4 Events-chain concept maps show the order of steps in a process or event. This concept map shows how a sound makes an echo.

Cycle Map A specific type of events chain is a cycle map. It is used when the series of events do not produce a final outcome, but instead relate back to the beginning event, such as in **Figure 5.** Therefore, the cycle repeats itself.

To make a cycle map, first decide what event is the beginning event. This is also called the initiating event. Then list the next events in the order that they occur, with the last event relating back to the initiating event. Words can be written between the events that describe what happens from one event to the next. The number of events in a cycle map can vary, but usually contain three or more events.

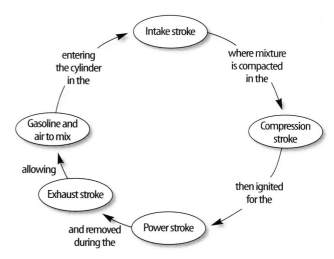

Figure 5 A cycle map shows events that occur in a cycle.

Spider Map A type of concept map that you can use for brainstorming is the spider map. When you have a central idea, you might find that you have a jumble of ideas that relate to it but are not necessarily clearly related to each other. The spider map on sound in **Figure 6** shows that if you write these ideas outside the main concept, then you can begin to separate and group unrelated terms so they become more useful.

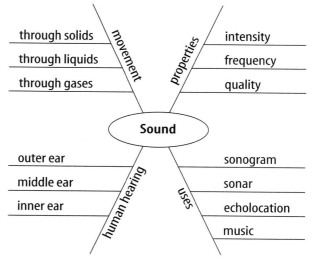

Figure 6 A spider map allows you to list ideas that relate to a central topic but not necessarily to one another.

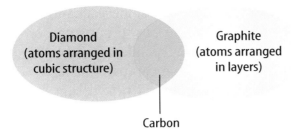

Figure 7 This Venn diagram compares and contrasts two substances made from carbon.

Venn Diagram To illustrate how two subjects compare and contrast you can use a Venn diagram. You can see the characteristics that the subjects have in common and those that they do not, shown in **Figure 7.**

To create a Venn diagram, draw two overlapping ovals that that are big enough to write in. List the characteristics unique to one subject in one oval, and the characteristics of the other subject in the other oval. The characteristics in common are listed in the overlapping section.

Make and Use Tables One way to organize information so it is easier to understand is to use a table. Tables can contain numbers, words, or both.

To make a table, list the items to be compared in the first column and the characteristics to be compared in the first row. The title should clearly indicate the content of the table, and the column or row heads should be clear. Notice that in **Table 1** the units are included.

Table 1 Recyclables Collected During Week			
Day of Week	Paper (kg)	Aluminum (kg)	Glass (kg)
Monday	5.0	4.0	12.0
Wednesday	4.0	1.0	10.0
Friday	2.5	2.0	10.0

Make a Model One way to help you better understand the parts of a structure, the way a process works, or to show things too large or small for viewing is to make a model. For example, an atomic model made of a plastic-ball nucleus and pipe-cleaner electron shells can help you visualize how the parts of an atom relate to each other. Other types of models can by devised on a computer or represented by equations.

Form a Hypothesis

A possible explanation based on previous knowledge and observations is called a hypothesis. After researching gasoline types and recalling previous experiences in your family's car you form a hypothesis—our car runs more efficiently because we use premium gasoline. To be valid, a hypothesis has to be something you can test by using an investigation.

Predict When you apply a hypothesis to a specific situation, you predict something about that situation. A prediction makes a statement in advance, based on prior observation, experience, or scientific reasoning. People use predictions to make everyday decisions. Scientists test predictions by performing investigations. Based on previous observations and experiences, you might form a prediction that cars are more efficient with premium gasoline. The prediction can be tested in an investigation.

Design an Experiment A scientist needs to make many decisions before beginning an investigation. Some of these include: how to carry out the investigation, what steps to follow, how to record the data, and how the investigation will answer the question. It also is important to address any safety concerns.

Test the Hypothesis

Now that you have formed your hypothesis, you need to test it. Using an investigation, you will make observations and collect data, or information. This data might either support or not support your hypothesis. Scientists collect and organize data as numbers and descriptions.

Follow a Procedure In order to know what materials to use, as well as how and in what order to use them, you must follow a procedure. **Figure 8** shows a procedure you might follow to test your hypothesis.

Procedure
1. Use regular gasoline for two weeks.
2. Record the number of kilometers between fill-ups and the amount of gasoline used.
3. Switch to premium gasoline for two weeks.
4. Record the number of kilometers between fill-ups and the amount of gasoline used.

Figure 8 A procedure tells you what to do step by step.

Identify and Manipulate Variables and Controls In any experiment, it is important to keep everything the same except for the item you are testing. The one factor you change is called the independent variable. The change that results is the dependent variable. Make sure you have only one independent variable, to assure yourself of the cause of the changes you observe in the dependent variable. For example, in your gasoline experiment the type of fuel is the independent variable. The dependent variable is the efficiency.

Many experiments also have a control— an individual instance or experimental subject for which the independent variable is not changed. You can then compare the test results to the control results. To design a control you can have two cars of the same type. The control car uses regular gasoline for four weeks. After you are done with the test, you can compare the experimental results to the control results.

Collect Data

Whether you are carrying out an investigation or a short observational experiment, you will collect data, as shown in **Figure 9.** Scientists collect data as numbers and descriptions and organize it in specific ways.

Observe Scientists observe items and events, then record what they see. When they use only words to describe an observation, it is called qualitative data. Scientists' observations also can describe how much there is of something. These observations use numbers, as well as words, in the description and are called quantitative data. For example, if a sample of the element gold is described as being "shiny and very dense" the data are qualitative. Quantitative data on this sample of gold might include "a mass of 30 g and a density of 19.3 g/cm^3."

Figure 9 Collecting data is one way to gather information directly.

Figure 10 Record data neatly and clearly so it is easy to understand.

When you make observations you should examine the entire object or situation first, and then look carefully for details. It is important to record observations accurately and completely. Always record your notes immediately as you make them, so you do not miss details or make a mistake when recording results from memory. Never put unidentified observations on scraps of paper. Instead they should be recorded in a notebook, like the one in **Figure 10.** Write your data neatly so you can easily read it later. At each point in the experiment, record your observations and label them. That way, you will not have to determine what the figures mean when you look at your notes later. Set up any tables that you will need to use ahead of time, so you can record any observations right away. Remember to avoid bias when collecting data by not including personal thoughts when you record observations. Record only what you observe.

Estimate Scientific work also involves estimating. To estimate is to make a judgment about the size or the number of something without measuring or counting. This is important when the number or size of an object or population is too large or too difficult to accurately count or measure.

Sample Scientists may use a sample or a portion of the total number as a type of estimation. To sample is to take a small, representative portion of the objects or organisms of a population for research. By making careful observations or manipulating variables within that portion of the group, information is discovered and conclusions are drawn that might apply to the whole population. A poorly chosen sample can be unrepresentative of the whole. If you were trying to determine the rainfall in an area, it would not be best to take a rainfall sample from under a tree.

Measure You use measurements everyday. Scientists also take measurements when collecting data. When taking measurements, it is important to know how to use measuring tools properly. Accuracy also is important.

Length To measure length, the distance between two points, scientists use meters. Smaller measurements might be measured in centimeters or millimeters.

Length is measured using a metric ruler or meter stick. When using a metric ruler, line up the 0-cm mark with the end of the object being measured and read the number of the unit where the object ends. Look at the metric ruler shown in **Figure 11.** The centimeter lines are the long, numbered lines, and the shorter lines are millimeter lines. In this instance, the length would be 4.50 cm.

Figure 11 This metric ruler has centimeter and millimeter divisions.

Mass The SI unit for mass is the kilogram (kg). Scientists can measure mass using units formed by adding metric prefixes to the unit gram (g), such as milligram (mg). To measure mass, you might use a triple-beam balance similar to the one shown in **Figure 12.** The balance has a pan on one side and a set of beams on the other side. Each beam has a rider that slides on the beam.

When using a triple-beam balance, place an object on the pan. Slide the largest rider along its beam until the pointer drops below zero. Then move it back one notch. Repeat the process for each rider proceeding from the larger to smaller until the pointer swings an equal distance above and below the zero point. Sum the masses on each beam to find the mass of the object. Move all riders back to zero when finished.

Instead of putting materials directly on the balance, scientists often take a tare of a container. A tare is the mass of a container into which objects or substances are placed for measuring their masses. To mass objects or substances, find the mass of a clean container. Remove the container from the pan, and place the object or substances in the container. Find the mass of the container with the materials in it. Subtract the mass of the empty container from the mass of the filled container to find the mass of the materials you are using.

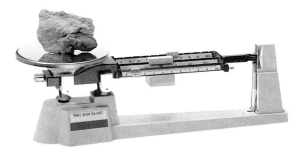

Figure 12 A triple-beam balance is used to determine the mass of an object.

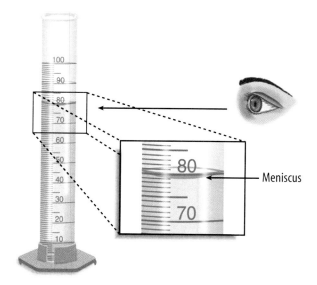

Meniscus

Figure 13 Graduated cylinders measure liquid volume.

Liquid Volume To measure liquids, the unit used is the liter. When a smaller unit is needed, scientists might use a milliliter. Because a milliliter takes up the volume of a cube measuring 1 cm on each side it also can be called a cubic centimeter ($cm^3 = cm \times cm \times cm$).

You can use beakers and graduated cylinders to measure liquid volume. A graduated cylinder, shown in **Figure 13,** is marked from bottom to top in milliliters. In lab, you might use a 10-mL graduated cylinder or a 100-mL graduated cylinder. When measuring liquids, notice that the liquid has a curved surface. Look at the surface at eye level, and measure the bottom of the curve. This is called the meniscus. The graduated cylinder in **Figure 13** contains 79.0 mL, or 79.0 cm^3, of a liquid.

Temperature Scientists often measure temperature using the Celsius scale. Pure water has a freezing point of 0°C and boiling point of 100°C. The unit of measurement is degrees Celsius. Two other scales often used are the Fahrenheit and Kelvin scales.

Figure 14 A thermometer measures the temperature of an object.

Scientists use a thermometer to measure temperature. Most thermometers in a laboratory are glass tubes with a bulb at the bottom end containing a liquid such as colored alcohol. The liquid rises or falls with a change in temperature. To read a glass thermometer like the thermometer in **Figure 14,** rotate it slowly until a red line appears. Read the temperature where the red line ends.

Form Operational Definitions An operational definition defines an object by how it functions, works, or behaves. For example, when you are playing hide and seek and a tree is home base, you have created an operational definition for a tree.

Objects can have more than one operational definition. For example, a ruler can be defined as a tool that measures the length of an object (how it is used). It can also be a tool with a series of marks used as a standard when measuring (how it works).

Analyze the Data

To determine the meaning of your observations and investigation results, you will need to look for patterns in the data. Then you must think critically to determine what the data mean. Scientists use several approaches when they analyze the data they have collected and recorded. Each approach is useful for identifying specific patterns.

Interpret Data The word *interpret* means "to explain the meaning of something." When analyzing data from an experiement, try to find out what the data show. Identify the control group and the test group to see whether or not changes in the independent variable have had an effect. Look for differences in the dependent variable between the control and test groups.

Classify Sorting objects or events into groups based on common features is called classifying. When classifying, first observe the objects or events to be classified. Then select one feature that is shared by some members in the group, but not by all. Place those members that share that feature in a subgroup. You can classify members into smaller and smaller subgroups based on characteristics. Remember that when you classify, you are grouping objects or events for a purpose. Keep your purpose in mind as you select the features to form groups and subgroups.

Compare and Contrast Observations can be analyzed by noting the similarities and differences between two more objects or events that you observe. When you look at objects or events to see how they are similar, you are comparing them. Contrasting is looking for differences in objects or events.

Recognize Cause and Effect A cause is a reason for an action or condition. The effect is that action or condition. When two events happen together, it is not necessarily true that one event caused the other. Scientists must design a controlled investigation to recognize the exact cause and effect.

Draw Conclusions

When scientists have analyzed the data they collected, they proceed to draw conclusions about the data. These conclusions are sometimes stated in words similar to the hypothesis that you formed earlier. They may confirm a hypothesis, or lead you to a new hypothesis.

Infer Scientists often make inferences based on their observations. An inference is an attempt to explain observations or to indicate a cause. An inference is not a fact, but a logical conclusion that needs further investigation. For example, you may infer that a fire has caused smoke. Until you investigate, however, you do not know for sure.

Apply When you draw a conclusion, you must apply those conclusions to determine whether the data supports the hypothesis. If your data do not support your hypothesis, it does not mean that the hypothesis is wrong. It means only that the result of the investigation did not support the hypothesis. Maybe the experiment needs to be redesigned, or some of the initial observations on which the hypothesis was based were incomplete or biased. Perhaps more observation or research is needed to refine your hypothesis. A successful investigation does not always come out the way you originally predicted.

Avoid Bias Sometimes a scientific investigation involves making judgments. When you make a judgment, you form an opinion. It is important to be honest and not to allow any expectations of results to bias your judgments. This is important throughout the entire investigation, from researching to collecting data to drawing conclusions.

Communicate

The communication of ideas is an important part of the work of scientists. A discovery that is not reported will not advance the scientific community's understanding or knowledge. Communication among scientists also is important as a way of improving their investigations.

Scientists communicate in many ways, from writing articles in journals and magazines that explain their investigations and experiments, to announcing important discoveries on television and radio. Scientists also share ideas with colleagues on the Internet or present them as lectures, like the student is doing in **Figure 15.**

Figure 15 A student communicates to his peers about his investigation.

SAFETY SYMBOLS

SAFETY SYMBOLS		HAZARD	EXAMPLES	PRECAUTION	REMEDY
DISPOSAL		Special disposal procedures need to be followed.	certain chemicals, living organisms	Do not dispose of these materials in the sink or trash can.	Dispose of wastes as directed by your teacher.
BIOLOGICAL		Organisms or other biological materials that might be harmful to humans	bacteria, fungi, blood, unpreserved tissues, plant materials	Avoid skin contact with these materials. Wear mask or gloves.	Notify your teacher if you suspect contact with material. Wash hands thoroughly.
EXTREME TEMPERATURE		Objects that can burn skin by being too cold or too hot	boiling liquids, hot plates, dry ice, liquid nitrogen	Use proper protection when handling.	Go to your teacher for first aid.
SHARP OBJECT		Use of tools or glassware that can easily puncture or slice skin	razor blades, pins, scalpels, pointed tools, dissecting probes, broken glass	Practice common-sense behavior and follow guidelines for use of the tool.	Go to your teacher for first aid.
FUME		Possible danger to respiratory tract from fumes	ammonia, acetone, nail polish remover, heated sulfur, moth balls	Make sure there is good ventilation. Never smell fumes directly. Wear a mask.	Leave foul area and notify your teacher immediately.
ELECTRICAL		Possible danger from electrical shock or burn	improper grounding, liquid spills, short circuits, exposed wires	Double-check setup with teacher. Check condition of wires and apparatus.	Do not attempt to fix electrical problems. Notify your teacher immediately.
IRRITANT		Substances that can irritate the skin or mucous membranes of the respiratory tract	pollen, moth balls, steel wool, fiberglass, potassium permanganate	Wear dust mask and gloves. Practice extra care when handling these materials.	Go to your teacher for first aid.
CHEMICAL		Chemicals can react with and destroy tissue and other materials	bleaches such as hydrogen peroxide; acids such as sulfuric acid, hydrochloric acid; bases such as ammonia, sodium hydroxide	Wear goggles, gloves, and an apron.	Immediately flush the affected area with water and notify your teacher.
TOXIC		Substance may be poisonous if touched, inhaled, or swallowed.	mercury, many metal compounds, iodine, poinsettia plant parts	Follow your teacher's instructions.	Always wash hands thoroughly after use. Go to your teacher for first aid.
FLAMMABLE		Flammable chemicals may be ignited by open flame, spark, or exposed heat.	alcohol, kerosene, potassium permanganate	Avoid open flames and heat when using flammable chemicals.	Notify your teacher immediately. Use fire safety equipment if applicable.
OPEN FLAME		Open flame in use, may cause fire.	hair, clothing, paper, synthetic materials	Tie back hair and loose clothing. Follow teacher's instruction on lighting and extinguishing flames.	Notify your teacher immediately. Use fire safety equipment if applicable.

 Eye Safety Proper eye protection should be worn at all times by anyone performing or observing science activities.

 Clothing Protection This symbol appears when substances could stain or burn clothing.

 Animal Safety This symbol appears when safety of animals and students must be ensured.

 Handwashing After the lab, wash hands with soap and water before removing goggles.

Safety in the Science Laboratory

The science laboratory is a safe place to work if you follow standard safety procedures. Being responsible for your own safety helps to make the entire laboratory a safer place for everyone. When performing any lab, read and apply the caution statements and safety symbol listed at the beginning of the lab.

General Safety Rules

1. Obtain your teacher's permission to begin all investigations and use laboratory equipment.

2. Study the procedure. Ask your teacher any questions. Be sure you understand safety symbols shown on the page.

3. Notify your teacher about allergies or other health conditions which can affect your participation in a lab.

4. Learn and follow use and safety procedures for your equipment. If unsure, ask your teacher.

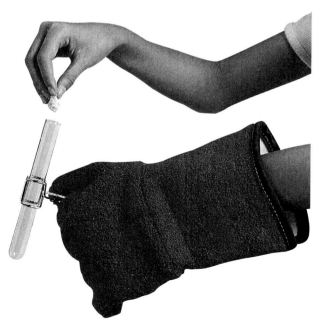

5. Never eat, drink, chew gum, apply cosmetics, or do any personal grooming in the lab. Never use lab glassware as food or drink containers. Keep your hands away from your face and mouth.

6. Know the location and proper use of the safety shower, eye wash, fire blanket, and fire alarm.

Prevent Accidents

1. Use the safety equipment provided to you. Goggles and a safety apron should be worn during investigations.

2. Do NOT use hair spray, mousse, or other flammable hair products. Tie back long hair and tie down loose clothing.

3. Do NOT wear sandals or other open-toed shoes in the lab.

4. Remove jewelry on hands and wrists. Loose jewelry, such as chains and long necklaces, should be removed to prevent them from getting caught in equipment.

5. Do not taste any substances or draw any material into a tube with your mouth.

6. Proper behavior is expected in the lab. Practical jokes and fooling around can lead to accidents and injury.

7. Keep your work area uncluttered.

Laboratory Work

1. Collect and carry all equipment and materials to your work area before beginning a lab.

2. Remain in your own work area unless given permission by your teacher to leave it.

3. Always slant test tubes away from your-self and others when heating them, adding substances to them, or rinsing them.

4. If instructed to smell a substance in a container, hold the container a short dis-tance away and fan vapors towards your nose.

5. Do NOT substitute other chemicals/sub-stances for those in the materials list unless instructed to do so by your teacher.

6. Do NOT take any materials or chemicals outside of the laboratory.

7. Stay out of storage areas unless instructed to be there and supervised by your teacher.

Laboratory Cleanup

1. Turn off all burners, water, and gas, and disconnect all electrical devices.

2. Clean all pieces of equipment and return all materials to their proper places.

3. Dispose of chemicals and other materi-als as directed by your teacher. Place broken glass and solid substances in the proper containers. Never discard mate-rials in the sink.

4. Clean your work area.

5. Wash your hands with soap and water thoroughly BEFORE removing your goggles.

Emergencies

1. Report any fire, electrical shock, glass-ware breakage, spill, or injury, no matter how small, to your teacher immediately. Follow his or her instructions.

2. If your clothing should catch fire, STOP, DROP, and ROLL. If possible, smother it with the fire blanket or get under a safety shower. NEVER RUN.

3. If a fire should occur, turn off all gas and leave the room according to established procedures.

4. In most instances, your teacher will clean up spills. Do NOT attempt to clean up spills unless you are given per-mission and instructions to do so.

5. If chemicals come into contact with your eyes or skin, notify your teacher immedi-ately. Use the eyewash or flush your skin or eyes with large quantities of water.

6. The fire extinguisher and first-aid kit should only be used by your teacher unless it is an extreme emergency and you have been given permission.

7. If someone is injured or becomes ill, only a professional medical provider or some-one certified in first aid should perform first-aid procedures.

EXTRA  Labs

From Your Kitchen, Junk Drawer, or Yard

1 The Pressure's On

Real-World Question

How can atmospheric air pressure changes be modeled?

Possible Materials
- large pot
- stove or hotplate
- tongs
- oven mitt
- empty aluminum soda can
- water
- cold water
- small aquarium or large bowl
- large jar
- measuring cup

Procedure

1. Fill a small aquarium or large bowl with cold water.
2. Pour water into a large pot and boil it.
3. Pour 25 mL of water into an empty aluminum can.
4. Using an oven mitt and tongs, hold the bottom of the can in the boiling water for 1 min.
5. Remove the can from the pot and immediately submerge it upside-down in the cold water in the aquarium or large bowl.

Conclude and Apply

1. Describe what happened to the can in the cold water.
2. Infer why the can changed in the cold water.

2 Bottling a Tornado

Real-World Question

How can you model a tornado?

Possible Materials
- 2-L soda bottles (2)
- dish soap
- masking tape
- duct tape
- measuring cup
- towel

Procedure

1. Remove the labels from two 2-L soda bottles.
2. Fill one bottle with 1.5 L of water.
3. Add two drops of dish soap to the bottle with the water.

4. Invert the second bottle and connect the openings of the bottles.
5. Attach the two bottles together with duct tape.
6. Flip the bottles upside-down and quickly swirl the top bottle with a smooth motion. Observe the tornado pattern made in the water.

Conclude and Apply

1. Describe how you modeled a tornado.
2. Research how a real tornado forms.

③ Getting Warmer

▶ Real-World Question
How do different surfaces affect temperature?

Possible Materials 🔌
- thermometer
- moist leaf litter
- large self-sealing bag
- stopwatch or watch

▶ Procedure
1. Collect moist leaf litter in a large self-sealing bag.
2. Pile the leaf litter on a patch of grass that is exposed to direct sunlight.

3. Set the thermometer in the center of the leaf litter, wait 3 min, and measure the temperature.
4. Place the thermometer on the grass in direct sunlight, wait 3 min, and measure the temperature.
5. Place the thermometer on a cement surface in direct sunlight, wait 3 min, and measure the temperature.
6. Place the thermometer on an asphalt surface in direct sunlight, wait 3 min, and measure the temperature.

▶ Conclude and Apply
1. Compare the temperatures of the different surfaces.
2. Explain how you measured the heat-island effect.

④ Making Smog

▶ Real-World Question
What does air pollution look like up close?

Possible Materials 🔲 🔥 🥽 🧤 🔌
- small glass jar
- black construction paper
- large, wide mouth glass jar
- small aquarium or large ceramic or glass bowl
- small wooden blocks (4)
- matches
- oven mitts

▶ Procedure
1. Take all your materials outside.
2. Turn a small aquarium upside down and place the four wood blocks under the edges so that it stands 3–4 cm off the ground.

3. Fold the black construction paper in half five times but leave it loosely folded.
4. Light the top of the construction paper with a match, and immediately insert the paper into the small jar.
5. Place the small jar under the aquarium and observe the air pollution created by the burning paper.

▶ Conclude and Apply
1. Describe how this activity models air pollution.
2. Infer how you could have reduced the air pollution caused by the burning paper.

Computer Skills

People who study science rely on computers, like the one in **Figure 16,** to record and store data and to analyze results from investigations. Whether you work in a laboratory or just need to write a lab report with tables, good computer skills are a necessity.

Using the computer comes with responsibility. Issues of ownership, security, and privacy can arise. Remember, if you did not author the information you are using, you must provide a source for your information. Also, anything on a computer can be accessed by others. Do not put anything on the computer that you would not want everyone to know. To add more security to your work, use a password.

Use a Word Processing Program

A computer program that allows you to type your information, change it as many times as you need to, and then print it out is called a word processing program. Word processing programs also can be used to make tables.

Figure 16 A computer will make reports neater and more professional looking.

Learn the Skill To start your word processing program, a blank document, sometimes called "Document 1," appears on the screen. To begin, start typing. To create a new document, click the *New* button on the standard tool bar. These tips will help you format the document.

- The program will automatically move to the next line; press *Enter* if you wish to start a new paragraph.
- Symbols, called non-printing characters, can be hidden by clicking the *Show/Hide* button on your toolbar.
- To insert text, move the cursor to the point where you want the insertion to go, click on the mouse once, and type the text.
- To move several lines of text, select the text and click the *Cut* button on your toolbar. Then position your cursor in the location that you want to move the cut text and click *Paste.* If you move to the wrong place, click *Undo.*
- The spell check feature does not catch words that are misspelled to look like other words, like "cold" instead of "gold." Always reread your document to catch all spelling mistakes.
- To learn about other word processing methods, read the user's manual or click on the *Help* button.
- You can integrate databases, graphics, and spreadsheets into documents by copying from another program and pasting it into your document, or by using desktop publishing (DTP). DTP software allows you to put text and graphics together to finish your document with a professional look. This software varies in how it is used and its capabilities.

Use a Database

A collection of facts stored in a computer and sorted into different fields is called a database. A database can be reorganized in any way that suits your needs.

Learn the Skill A computer program that allows you to create your own database is a database management system (DBMS). It allows you to add, delete, or change information. Take time to get to know the features of your database software.

- Determine what facts you would like to include and research to collect your information.
- Determine how you want to organize the information.
- Follow the instructions for your particular DBMS to set up fields. Then enter each item of data in the appropriate field.
- Follow the instructions to sort the information in order of importance.
- Evaluate the information in your database, and add, delete, or change as necessary.

Use the Internet

The Internet is a global network of computers where information is stored and shared. To use the Internet, like the students in **Figure 17,** you need a modem to connect your computer to a phone line and an Internet Service Provider account.

Learn the Skill To access internet sites and information, use a "Web browser," which lets you view and explore pages on the World Wide Web. Each page is its own site, and each site has its own address, called a URL. Once you have found a Web browser, follow these steps for a search (this also is how you search a database).

Figure 17 The Internet allows you to search a global network for a variety of information.

- Be as specific as possible. If you know you want to research "gold," don't type in "elements." Keep narrowing your search until you find what you want.
- Web sites that end in *.com* are commercial Web sites; *.org, .edu,* and *.gov* are non-profit, educational, or government Web sites.
- Electronic encyclopedias, almanacs, indexes, and catalogs will help locate and select relevant information.
- Develop a "home page" with relative ease. When developing a Web site, NEVER post pictures or disclose personal information such as location, names, or phone numbers. Your school or community usually can host your Web site. A basic understanding of HTML (hypertext mark-up language), the language of Web sites, is necessary. Software that creates HTML code is called authoring software, and can be downloaded free from many Web sites. This software allows text and pictures to be arranged as the software is writing the HTML code.

Use a Spreadsheet

A spreadsheet, shown in **Figure 18,** can perform mathematical functions with any data arranged in columns and rows. By entering a simple equation into a cell, the program can perform operations in specific cells, rows, or columns.

Learn the Skill Each column (vertical) is assigned a letter, and each row (horizontal) is assigned a number. Each point where a row and column intersect is called a cell, and is labeled according to where it is located—Column A, Row 1 (A1).

■ Decide how to organize the data, and enter it in the correct row or column.
■ Spreadsheets can use standard formulas or formulas can be customized to calculate cells.
■ To make a change, click on a cell to make it activate, and enter the edited data or formula.
■ Spreadsheets also can display your results in graphs. Choose the style of graph that best represents the data.

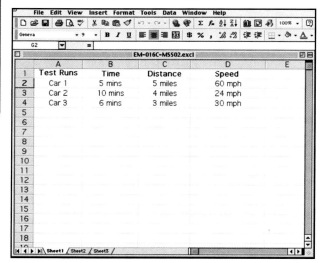

Figure 18 A spreadsheet allows you to perform mathematical operations on your data.

Use Graphics Software

Adding pictures, called graphics, to your documents is one way to make your documents more meaningful and exciting. This software adds, edits, and even constructs graphics. There is a variety of graphics software programs. The tools used for drawing can be a mouse, keyboard, or other specialized devices. Some graphics programs are simple. Others are complicated, called computer-aided design (CAD) software.

Learn the Skill It is important to have an understanding of the graphics software being used before starting. The better the software is understood, the better the results. The graphics can be placed in a word-processing document.

■ Clip art can be found on a variety of internet sites, and on CDs. These images can be copied and pasted into your document.
■ When beginning, try editing existing drawings, then work up to creating drawings.
■ The images are made of tiny rectangles of color called pixels. Each pixel can be altered.
■ Digital photography is another way to add images. The photographs in the memory of a digital camera can be downloaded into a computer, then edited and added to the document.
■ Graphics software also can allow animation. The software allows drawings to have the appearance of movement by connecting basic drawings automatically. This is called in-betweening, or tweening.
■ Remember to save often.

Presentation Skills

Develop Multimedia Presentations

Most presentations are more dynamic if they include diagrams, photographs, videos, or sound recordings, like the one shown in **Figure 19.** A multimedia presentation involves using stereos, overhead projectors, televisions, computers, and more.

Learn the Skill Decide the main points of your presentation, and what types of media would best illustrate those points.

- Make sure you know how to use the equipment you are working with.
- Practice the presentation using the equipment several times.
- Enlist the help of a classmate to push play or turn lights out for you. Be sure to practice your presentation with him or her.
- If possible, set up all of the equipment ahead of time, and make sure everything is working properly.

Figure 19 These students are engaging the audience using a variety of tools.

Computer Presentations

There are many different interactive computer programs that you can use to enhance your presentation. Most computers have a compact disc (CD) drive that can play both CDs and digital video discs (DVDs). Also, there is hardware to connect a regular CD, DVD, or VCR. These tools will enhance your presentation.

Another method of using the computer to aid in your presentation is to develop a slide show using a computer program. This can allow movement of visuals at the presenter's pace, and can allow for visuals to build on one another.

Learn the Skill In order to create multimedia presentations on a computer, you need to have certain tools. These may include traditional graphic tools and drawing programs, animation programs, and authoring systems that tie everything together. Your computer will tell you which tools it supports. The most important step is to learn about the tools that you will be using.

- Often, color and strong images will convey a point better than words alone. Use the best methods available to convey your point.
- As with other presentations, practice many times.
- Practice your presentation with the tools you and any assistants will be using.
- Maintain eye contact with the audience. The purpose of using the computer is not to prompt the presenter, but to help the audience understand the points of the presentation.

Math Review

Use Fractions

A fraction compares a part to a whole. In the fraction $\frac{2}{3}$, the 2 represents the part and is the numerator. The 3 represents the whole and is the denominator.

Reduce Fractions To reduce a fraction, you must find the largest factor that is common to both the numerator and the denominator, the greatest common factor (GCF). Divide both numbers by the GCF. The fraction has then been reduced, or it is in its simplest form.

Example Twelve of the 20 chemicals in the science lab are in powder form. What fraction of the chemicals used in the lab are in powder form?

Step 1 Write the fraction.

$$\frac{\text{part}}{\text{whole}} = \frac{12}{20}$$

Step 2 To find the GCF of the numerator and denominator, list all of the factors of each number.

Factors of 12: 1, 2, 3, 4, 6, 12 (the numbers that divide evenly into 12)

Factors of 20: 1, 2, 4, 5, 10, 20 (the numbers that divide evenly into 20)

Step 3 List the common factors.

1, 2, 4.

Step 4 Choose the greatest factor in the list.

The GCF of 12 and 20 is 4.

Step 5 Divide the numerator and denominator by the GCF.

$$\frac{12 \div 4}{20 \div 4} = \frac{3}{5}$$

In the lab, $\frac{3}{5}$ of the chemicals are in powder form.

Practice Problem At an amusement park, 66 of 90 rides have a height restriction. What fraction of the rides, in its simplest form, has a height restriction?

Add and Subtract Fractions To add or subtract fractions with the same denominator, add or subtract the numerators and write the sum or difference over the denominator. After finding the sum or difference, find the simplest form for your fraction.

Example 1 In the forest outside your house, $\frac{1}{8}$ of the animals are rabbits, $\frac{3}{8}$ are squirrels, and the remainder are birds and insects. How many are mammals?

Step 1 Add the numerators.

$$\frac{1}{8} + \frac{3}{8} = \frac{(1 + 3)}{8} = \frac{4}{8}$$

Step 2 Find the GCF.

$$\frac{4}{8} \quad \text{(GCF, 4)}$$

Step 3 Divide the numerator and denominator by the GCF.

$$\frac{4}{4} = 1, \quad \frac{8}{4} = 2$$

$\frac{1}{2}$ of the animals are mammals.

Example 2 If $\frac{7}{16}$ of the Earth is covered by freshwater, and $\frac{1}{16}$ of that is in glaciers, how much freshwater is not frozen?

Step 1 Subtract the numerators.

$$\frac{7}{16} - \frac{1}{16} = \frac{(7 - 1)}{16} = \frac{6}{16}$$

Step 2 Find the GCF.

$$\frac{6}{16} \quad \text{(GCF, 2)}$$

Step 3 Divide the numerator and denominator by the GCF.

$$\frac{6}{2} = 3, \quad \frac{16}{2} = 8$$

$\frac{3}{8}$ of the freshwater is not frozen.

Practice Problem A bicycle rider is going 15 km/h for $\frac{4}{9}$ of his ride, 10 km/h for $\frac{2}{9}$ of his ride, and 8 km/h for the remainder of the ride. How much of his ride is he going over 8 km/h?

Unlike Denominators To add or subtract fractions with unlike denominators, first find the least common denominator (LCD). This is the smallest number that is a common multiple of both denominators. Rename each fraction with the LCD, and then add or subtract. Find the simplest form if necessary.

Example 1 A chemist makes a paste that is $\frac{1}{2}$ table salt (NaCl), $\frac{1}{3}$ sugar ($C_6H_{12}O_6$), and the rest water (H_2O). How much of the paste is a solid?

Step 1 Find the LCD of the fractions.
$$\frac{1}{2} + \frac{1}{3} \quad (LCD, 6)$$

Step 2 Rename each numerator and each denominator with the LCD.
$$1 \times 3 = 3, \quad 2 \times 3 = 6$$
$$1 \times 2 = 2, \quad 3 \times 2 = 6$$

Step 3 Add the numerators.
$$\frac{3}{6} + \frac{2}{6} = \frac{(3 + 2)}{6} = \frac{5}{6}$$

$\frac{5}{6}$ of the paste is a solid.

Example 2 The average precipitation in Grand Junction, CO, is $\frac{7}{10}$ inch in November, and $\frac{3}{5}$ inch in December. What is the total average precipitation?

Step 1 Find the LCD of the fractions.
$$\frac{7}{10} + \frac{3}{5} \quad (LCD, 10)$$

Step 2 Rename each numerator and each denominator with the LCD.
$$7 \times 1 = 7, \quad 10 \times 1 = 10$$
$$3 \times 2 = 6, \quad 5 \times 2 = 10$$

Step 3 Add the numerators.
$$\frac{7}{10} + \frac{6}{10} = \frac{(7 + 6)}{10} = \frac{13}{10}$$

$\frac{13}{10}$ inches total precipitation, or $1\frac{3}{10}$ inches.

Practice Problem On an electric bill, about $\frac{1}{8}$ of the energy is from solar energy and about $\frac{1}{10}$ is from wind power. How much of the total bill is from solar energy and wind power combined?

Example 3 In your body, $\frac{7}{10}$ of your muscle contractions are involuntary (cardiac and smooth muscle tissue). Smooth muscle makes $\frac{3}{15}$ of your muscle contractions. How many of your muscle contractions are made by cardiac muscle?

Step 1 Find the LCD of the fractions.
$$\frac{7}{10} - \frac{3}{15} \quad (LCD, 30)$$

Step 2 Rename each numerator and each denominator with the LCD.
$$7 \times 3 = 21, \quad 10 \times 3 = 30$$
$$3 \times 2 = 6, \quad 15 \times 2 = 30$$

Step 3 Subtract the numerators.
$$\frac{21}{30} - \frac{6}{30} = \frac{(21 - 6)}{30} = \frac{15}{30}$$

Step 4 Find the GCF.
$$\frac{15}{30} \quad (GCF, 15)$$
$$\frac{1}{2}$$

$\frac{1}{2}$ of all muscle contractions are cardiac muscle.

Example 4 Tony wants to make cookies that call for $\frac{3}{4}$ of a cup of flour, but he only has $\frac{1}{3}$ of a cup. How much more flour does he need?

Step 1 Find the LCD of the fractions.
$$\frac{3}{4} - \frac{1}{3} \quad (LCD, 12)$$

Step 2 Rename each numerator and each denominator with the LCD.
$$3 \times 3 = 9, \quad 4 \times 3 = 12$$
$$1 \times 4 = 4, \quad 3 \times 4 = 12$$

Step 3 Subtract the numerators.
$$\frac{9}{12} - \frac{4}{12} = \frac{(9 - 4)}{12} = \frac{5}{12}$$

$\frac{5}{12}$ of a cup of flour.

Practice Problem Using the information provided to you in Example 3 above, determine how many muscle contractions are voluntary (skeletal muscle).

Multiply Fractions To multiply with fractions, multiply the numerators and multiply the denominators. Find the simplest form if necessary.

Example Multiply $\frac{3}{5}$ by $\frac{1}{3}$.

Step 1 Multiply the numerators and denominators.
$$\frac{3}{5} \times \frac{1}{3} = \frac{(3 \times 1)}{(5 \times 3)} = \frac{3}{15}$$

Step 2 Find the GCF.
$$\frac{3}{15} \quad (\text{GCF, 3})$$

Step 3 Divide the numerator and denominator by the GCF.
$$\frac{3}{3} = 1, \ \frac{15}{3} = 5$$
$$\frac{1}{5}$$

$\frac{3}{5}$ multiplied by $\frac{1}{3}$ is $\frac{1}{5}$.

Practice Problem Multiply $\frac{3}{14}$ by $\frac{5}{16}$.

Find a Reciprocal Two numbers whose product is 1 are called multiplicative inverses, or reciprocals.

Example Find the reciprocal of $\frac{3}{8}$.

Step 1 Inverse the fraction by putting the denominator on top and the numerator on the bottom.
$$\frac{8}{3}$$

The reciprocal of $\frac{3}{8}$ is $\frac{8}{3}$.

Practice Problem Find the reciprocal of $\frac{4}{9}$.

Divide Fractions To divide one fraction by another fraction, multiply the dividend by the reciprocal of the divisor. Find the simplest form if necessary.

Example 1 Divide $\frac{1}{9}$ by $\frac{1}{3}$.

Step 1 Find the reciprocal of the divisor.
The reciprocal of $\frac{1}{3}$ is $\frac{3}{1}$.

Step 2 Multiply the dividend by the reciprocal of the divisor.
$$\frac{\frac{1}{9}}{\frac{1}{3}} = \frac{1}{9} \times \frac{3}{1} = \frac{(1 \times 3)}{(9 \times 1)} = \frac{3}{9}$$

Step 3 Find the GCF.
$$\frac{3}{9} \quad (\text{GCF, 3})$$

Step 4 Divide the numerator and denominator by the GCF.
$$\frac{3}{3} = 1, \ \frac{9}{3} = 3$$
$$\frac{1}{3}$$

$\frac{1}{9}$ divided by $\frac{1}{3}$ is $\frac{1}{3}$.

Example 2 Divide $\frac{3}{5}$ by $\frac{1}{4}$.

Step 1 Find the reciprocal of the divisor.
The reciprocal of $\frac{1}{4}$ is $\frac{4}{1}$.

Step 2 Multiply the dividend by the reciprocal of the divisor.
$$\frac{\frac{3}{5}}{\frac{1}{4}} = \frac{3}{5} \times \frac{4}{1} = \frac{(3 \times 4)}{(5 \times 1)} = \frac{12}{5}$$

$\frac{3}{5}$ divided by $\frac{1}{4}$ is $\frac{12}{5}$ or $2\frac{2}{5}$.

Practice Problem Divide $\frac{3}{11}$ by $\frac{7}{10}$.

Use Ratios

When you compare two numbers by division, you are using a ratio. Ratios can be written 3 to 5, 3:5, or $\frac{3}{5}$. Ratios, like fractions, also can be written in simplest form.

Ratios can represent probabilities, also called odds. This is a ratio that compares the number of ways a certain outcome occurs to the number of outcomes. For example, if you flip a coin 100 times, what are the odds that it will come up heads? There are two possible outcomes, heads or tails, so the odds of coming up heads are 50:100. Another way to say this is that 50 out of 100 times the coin will come up heads. In its simplest form, the ratio is 1:2.

Example 1 A chemical solution contains 40 g of salt and 64 g of baking soda. What is the ratio of salt to baking soda as a fraction in simplest form?

Step 1 Write the ratio as a fraction.
$$\frac{\text{salt}}{\text{baking soda}} = \frac{40}{64}$$

Step 2 Express the fraction in simplest form. The GCF of 40 and 64 is 8.
$$\frac{40}{64} = \frac{40 \div 8}{64 \div 8} = \frac{5}{8}$$

The ratio of salt to baking soda in the sample is 5:8.

Example 2 Sean rolls a 6-sided die 6 times. What are the odds that the side with a 3 will show?

Step 1 Write the ratio as a fraction.
$$\frac{\text{number of sides with a 3}}{\text{number of sides}} = \frac{1}{6}$$

Step 2 Multiply by the number of attempts.
$$\frac{1}{6} \times 6 \text{ attempts} = \frac{6}{6} \text{ attempts} = 1 \text{ attempt}$$

1 attempt out of 6 will show a 3.

Practice Problem Two metal rods measure 100 cm and 144 cm in length. What is the ratio of their lengths in simplest form?

Use Decimals

A fraction with a denominator that is a power of ten can be written as a decimal. For example, 0.27 means $\frac{27}{100}$. The decimal point separates the ones place from the tenths place.

Any fraction can be written as a decimal using division. For example, the fraction $\frac{5}{8}$ can be written as a decimal by dividing 5 by 8. Written as a decimal, it is 0.625.

Add or Subtract Decimals When adding and subtracting decimals, line up the decimal points before carrying out the operation.

Example 1 Find the sum of 47.68 and 7.80.

Step 1 Line up the decimal places when you write the numbers.
$$\begin{array}{r} 47.68 \\ +\ 7.80 \\ \hline \end{array}$$

Step 2 Add the decimals.
$$\begin{array}{r} 47.68 \\ +\ 7.80 \\ \hline 55.48 \end{array}$$

The sum of 47.68 and 7.80 is 55.48.

Example 2 Find the difference of 42.17 and 15.85.

Step 1 Line up the decimal places when you write the number.
$$\begin{array}{r} 42.17 \\ -15.85 \\ \hline \end{array}$$

Step 2 Subtract the decimals.
$$\begin{array}{r} 42.17 \\ -15.85 \\ \hline 26.32 \end{array}$$

The difference of 42.17 and 15.85 is 26.32.

Practice Problem Find the sum of 1.245 and 3.842.

Multiply Decimals To multiply decimals, multiply the numbers like any other number, ignoring the decimal point. Count the decimal places in each factor. The product will have the same number of decimal places as the sum of the decimal places in the factors.

Example Multiply 2.4 by 5.9.

Step 1 Multiply the factors like two whole numbers.

$24 \times 59 = 1416$

Step 2 Find the sum of the number of decimal places in the factors. Each factor has one decimal place, for a sum of two decimal places.

Step 3 The product will have two decimal places.

14.16

The product of 2.4 and 5.9 is 14.16.

Practice Problem Multiply 4.6 by 2.2.

Divide Decimals When dividing decimals, change the divisor to a whole number. To do this, multiply both the divisor and the dividend by the same power of ten. Then place the decimal point in the quotient directly above the decimal point in the dividend. Then divide as you do with whole numbers.

Example Divide 8.84 by 3.4.

Step 1 Multiply both factors by 10.

$3.4 \times 10 = 34$, $8.84 \times 10 = 88.4$

Step 2 Divide 88.4 by 34.

```
        2.6
  34)88.4
     −68
      204
     −204
        0
```

8.84 divided by 3.4 is 2.6.

Practice Problem Divide 75.6 by 3.6.

Use Proportions

An equation that shows that two ratios are equivalent is a proportion. The ratios $\frac{2}{4}$ and $\frac{5}{10}$ are equivalent, so they can be written as $\frac{2}{4} = \frac{5}{10}$. This equation is a proportion.

When two ratios form a proportion, the cross products are equal. To find the cross products in the proportion $\frac{2}{4} = \frac{5}{10}$, multiply the 2 and the 10, and the 4 and the 5. Therefore $2 \times 10 = 4 \times 5$, or $20 = 20$.

Because you know that both proportions are equal, you can use cross products to find a missing term in a proportion. This is known as solving the proportion.

Example The heights of a tree and a pole are proportional to the lengths of their shadows. The tree casts a shadow of 24 m when a 6-m pole casts a shadow of 4 m. What is the height of the tree?

Step 1 Write a proportion.

$$\frac{\text{height of tree}}{\text{height of pole}} = \frac{\text{length of tree's shadow}}{\text{length of pole's shadow}}$$

Step 2 Substitute the known values into the proportion. Let h represent the unknown value, the height of the tree.

$$\frac{h}{6} = \frac{24}{4}$$

Step 3 Find the cross products.

$h \times 4 = 6 \times 24$

Step 4 Simplify the equation.

$4h = 144$

Step 5 Divide each side by 4.

$$\frac{4h}{4} = \frac{144}{4}$$

$h = 36$

The height of the tree is 36 m.

Practice Problem The ratios of the weights of two objects on the Moon and on Earth are in proportion. A rock weighing 3 N on the Moon weighs 18 N on Earth. How much would a rock that weighs 5 N on the Moon weigh on Earth?

Math Skill Handbook

Use Percentages

The word *percent* means "out of one hundred." It is a ratio that compares a number to 100. Suppose you read that 77 percent of the Earth's surface is covered by water. That is the same as reading that the fraction of the Earth's surface covered by water is $\frac{77}{100}$. To express a fraction as a percent, first find the equivalent decimal for the fraction. Then, multiply the decimal by 100 and add the percent symbol.

Example Express $\frac{13}{20}$ as a percent.

Step 1 Find the equivalent decimal for the fraction.

$$\begin{array}{r} 0.65 \\ 20\overline{)13.00} \\ \underline{12\ 0} \\ 1\ 00 \\ \underline{1\ 00} \\ 0 \end{array}$$

Step 2 Rewrite the fraction $\frac{13}{20}$ as 0.65.

Step 3 Multiply 0.65 by 100 and add the % sign.
$0.65 \times 100 = 65 = 65\%$

So, $\frac{13}{20} = 65\%$.

This also can be solved as a proportion.

Example Express $\frac{13}{20}$ as a percent.

Step 1 Write a proportion.
$$\frac{13}{20} = \frac{x}{100}$$

Step 2 Find the cross products.
$1300 = 20x$

Step 3 Divide each side by 20.
$$\frac{1300}{20} = \frac{20x}{20}$$
$65\% = x$

Practice Problem In one year, 73 of 365 days were rainy in one city. What percent of the days in that city were rainy?

Solve One-Step Equations

A statement that two things are equal is an equation. For example, $A = B$ is an equation that states that A is equal to B.

An equation is solved when a variable is replaced with a value that makes both sides of the equation equal. To make both sides equal the inverse operation is used. Addition and subtraction are inverses, and multiplication and division are inverses.

Example 1 Solve the equation $x - 10 = 35$.

Step 1 Find the solution by adding 10 to each side of the equation.
$x - 10 = 35$
$x - 10 + 10 = 35 + 10$
$x = 45$

Step 2 Check the solution.
$x - 10 = 35$
$45 - 10 = 35$
$35 = 35$

Both sides of the equation are equal, so $x = 45$.

Example 2 In the formula $a = bc$, find the value of c if $a = 20$ and $b = 2$.

Step 1 Rearrange the formula so the unknown value is by itself on one side of the equation by dividing both sides by b.
$a = bc$
$\frac{a}{b} = \frac{bc}{b}$
$\frac{a}{b} = c$

Step 2 Replace the variables a and b with the values that are given.
$\frac{a}{b} = c$
$\frac{20}{2} = c$
$10 = c$

Step 3 Check the solution.
$a = bc$
$20 = 2 \times 10$
$20 = 20$

Both sides of the equation are equal, so $c = 10$ is the solution when $a = 20$ and $b = 2$.

Practice Problem In the formula $h = gd$, find the value of d if $g = 12.3$ and $h = 17.4$.

Use Statistics

The branch of mathematics that deals with collecting, analyzing, and presenting data is statistics. In statistics, there are three common ways to summarize data with a single number—the mean, the median, and the mode.

The **mean** of a set of data is the arithmetic average. It is found by adding the numbers in the data set and dividing by the number of items in the set.

The **median** is the middle number in a set of data when the data are arranged in numerical order. If there were an even number of data points, the median would be the mean of the two middle numbers.

The **mode** of a set of data is the number or item that appears most often.

Another number that often is used to describe a set of data is the range. The **range** is the difference between the largest number and the smallest number in a set of data.

A **frequency table** shows how many times each piece of data occurs, usually in a survey. **Table 2** below shows the results of a student survey on favorite color.

Table 2 Student Color Choice		
Color	Tally	Frequency
red	||||	4
blue	ⅢⅡ	5
black	||	2
green	|||	3
purple	ⅢⅡ ||	7
yellow	ⅢⅡ |	6

Based on the frequency table data, which color is the favorite?

Example The speeds (in m/s) for a race car during five different time trials are 39, 37, 44, 36, and 44.

To find the mean:

Step 1 Find the sum of the numbers.
$$39 + 37 + 44 + 36 + 44 = 200$$

Step 2 Divide the sum by the number of items, which is 5.
$$200 \div 5 = 40$$

The mean is 40 m/s.

To find the median:

Step 1 Arrange the measures from least to greatest.
36, 37, 39, 44, 44

Step 2 Determine the middle measure.
36, 37, <u>39</u>, 44, 44

The median is 39 m/s.

To find the mode:

Step 1 Group the numbers that are the same together.
44, 44, 36, 37, 39

Step 2 Determine the number that occurs most in the set.
<u>44, 44</u>, 36, 37, 39

The mode is 44 m/s.

To find the range:

Step 1 Arrange the measures from largest to smallest.
44, 44, 39, 37, 36

Step 2 Determine the largest and smallest measures in the set.
<u>44</u>, 44, 39, 37, <u>36</u>

Step 3 Find the difference between the largest and smallest measures.
$$44 - 36 = 8$$

The range is 8 m/s.

Practice Problem Find the mean, median, mode, and range for the data set 8, 4, 12, 8, 11, 14, 16.

Use Geometry

The branch of mathematics that deals with the measurement, properties, and relationships of points, lines, angles, surfaces, and solids is called geometry.

Perimeter The **perimeter** (P) is the distance around a geometric figure. To find the perimeter of a rectangle, add the length and width and multiply that sum by two, or $2(l + w)$. To find perimeters of irregular figures, add the length of the sides.

Example 1 Find the perimeter of a rectangle that is 3 m long and 5 m wide.

Step 1 You know that the perimeter is 2 times the sum of the width and length.
$P = 2(3 \text{ m} + 5 \text{ m})$

Step 2 Find the sum of the width and length.
$P = 2(8 \text{ m})$

Step 3 Multiply by 2.
$P = 16 \text{ m}$

The perimeter is 16 m.

Example 2 Find the perimeter of a shape with sides measuring 2 cm, 5 cm, 6 cm, 3 cm.

Step 1 You know that the perimeter is the sum of all the sides.
$P = 2 + 5 + 6 + 3$

Step 2 Find the sum of the sides.
$P = 2 + 5 + 6 + 3$
$P = 16$

The perimeter is 16 cm.

Practice Problem Find the perimeter of a rectangle with a length of 18 m and a width of 7 m.

Practice Problem Find the perimeter of a triangle measuring 1.6 cm by 2.4 cm by 2.4 cm.

Area of a Rectangle The **area** (A) is the number of square units needed to cover a surface. To find the area of a rectangle, multiply the length times the width, or $l \times w$. When finding area, the units also are multiplied. Area is given in square units.

Example Find the area of a rectangle with a length of 1 cm and a width of 10 cm.

Step 1 You know that the area is the length multiplied by the width.
$A = (1 \text{ cm} \times 10 \text{ cm})$

Step 2 Multiply the length by the width. Also multiply the units.
$A = 10 \text{ cm}^2$

The area is 10 cm².

Practice Problem Find the area of a square whose sides measure 4 m.

Area of a Triangle To find the area of a triangle, use the formula:

$$A = \frac{1}{2}(\text{base} \times \text{height})$$

The base of a triangle can be any of its sides. The height is the perpendicular distance from a base to the opposite endpoint, or vertex.

Example Find the area of a triangle with a base of 18 m and a height of 7 m.

Step 1 You know that the area is $\frac{1}{2}$ the base times the height.
$A = \frac{1}{2}(18 \text{ m} \times 7 \text{ m})$

Step 2 Multiply $\frac{1}{2}$ by the product of 18×7. Multiply the units.
$A = \frac{1}{2}(126 \text{ m}^2)$
$A = 63 \text{ m}^2$

The area is 63 m².

Practice Problem Find the area of a triangle with a base of 27 cm and a height of 17 cm.

Circumference of a Circle The **diameter** (*d*) of a circle is the distance across the circle through its center, and the **radius** (*r*) is the distance from the center to any point on the circle. The radius is half of the diameter. The distance around the circle is called the **circumference** (C). The formula for finding the circumference is:

$$C = 2\pi r \ \ or \ \ C = \pi d$$

The circumference divided by the diameter is always equal to 3.1415926... This nonterminating and nonrepeating number is represented by the Greek letter π (pi). An approximation often used for π is 3.14.

Example 1 Find the circumference of a circle with a radius of 3 m.

Step 1 You know the formula for the circumference is 2 times the radius times π.
$C = 2\pi(3)$

Step 2 Multiply 2 times the radius.
$C = 6\pi$

Step 3 Multiply by π.
$C = 19$ m

The circumference is 19 m.

Example 2 Find the circumference of a circle with a diameter of 24.0 cm.

Step 1 You know the formula for the circumference is the diameter times π.
$C = \pi(24.0)$

Step 2 Multiply the diameter by π.
$C = 75.4$ cm

The circumference is 75.4 cm.

Practice Problem Find the circumference of a circle with a radius of 19 cm.

Area of a Circle The formula for the area of a circle is:
$A = \pi r^2$

Example 1 Find the area of a circle with a radius of 4.0 cm.

Step 1 $A = \pi(4.0)^2$

Step 2 Find the square of the radius.
$A = 16\pi$

Step 3 Multiply the square of the radius by π.
$A = 50$ cm^2

The area of the circle is 50 cm^2.

Example 2 Find the area of a circle with a radius of 225 m.

Step 1 $A = \pi(225)^2$

Step 2 Find the square of the radius.
$A = 50625\pi$

Step 3 Multiply the square of the radius by π.
$A = 158962.5$

The area of the circle is 158,962 m^2.

Example 3 Find the area of a circle whose diameter is 20.0 mm.

Step 1 You know the formula for the area of a circle is the square of the radius times π, and that the radius is half of the diameter.
$A = \pi\left(\dfrac{20.0}{2}\right)^2$

Step 2 Find the radius.
$A = \pi(10.0)^2$

Step 3 Find the square of the radius.
$A = 100\pi$

Step 4 Multiply the square of the radius by π.
$A = 314$ mm^2

The area is 314 mm^2.

Practice Problem Find the area of a circle with a radius of 16 m.

Volume The measure of space occupied by a solid is the **volume** (V). To find the volume of a rectangular solid multiply the length times width times height, or $V = l \times w \times h$. It is measured in cubic units, such as cubic centimeters (cm^3).

Example Find the volume of a rectangular solid with a length of 2.0 m, a width of 4.0 m, and a height of 3.0 m.

Step 1 You know the formula for volume is the length times the width times the height.
$$V = 2.0 \text{ m} \times 4.0 \text{ m} \times 3.0 \text{ m}$$

Step 2 Multiply the length times the width times the height.
$$V = 24 \text{ m}^3$$

The volume is 24 m^3.

Practice Problem Find the volume of a rectangular solid that is 8 m long, 4 m wide, and 4 m high.

To find the volume of other solids, multiply the area of the base times the height.

Example 1 Find the volume of a solid that has a triangular base with a length of 8.0 m and a height of 7.0 m. The height of the entire solid is 15.0 m.

Step 1 You know that the base is a triangle, and the area of a triangle is $\frac{1}{2}$ the base times the height, and the volume is the area of the base times the height.
$$V = \left[\frac{1}{2}(b \times h)\right] \times 15$$

Step 2 Find the area of the base.
$$V = \left[\frac{1}{2}(8 \times 7)\right] \times 15$$
$$V = \left(\frac{1}{2} \times 56\right) \times 15$$

Step 3 Multiply the area of the base by the height of the solid.
$$V = 28 \times 15$$
$$V = 420 \text{ m}^3$$

The volume is 420 m^3.

Example 2 Find the volume of a cylinder that has a base with a radius of 12.0 cm, and a height of 21.0 cm.

Step 1 You know that the base is a circle, and the area of a circle is the square of the radius times π, and the volume is the area of the base times the height.
$$V = (\pi r^2) \times 21$$
$$V = (\pi 12^2) \times 21$$

Step 2 Find the area of the base.
$$V = 144\pi \times 21$$
$$V = 452 \times 21$$

Step 3 Multiply the area of the base by the height of the solid.
$$V = 9490 \text{ cm}^3$$

The volume is 9490 cm^3.

Example 3 Find the volume of a cylinder that has a diameter of 15 mm and a height of 4.8 mm.

Step 1 You know that the base is a circle with an area equal to the square of the radius times π. The radius is one-half the diameter. The volume is the area of the base times the height.
$$V = (\pi r^2) \times 4.8$$
$$V = \left[\pi\left(\frac{1}{2} \times 15\right)^2\right] \times 4.8$$
$$V = (\pi 7.5^2) \times 4.8$$

Step 2 Find the area of the base.
$$V = 56.25\pi \times 4.8$$
$$V = 176.63 \times 4.8$$

Step 3 Multiply the area of the base by the height of the solid.
$$V = 847.8$$

The volume is 847.8 mm^3.

Practice Problem Find the volume of a cylinder with a diameter of 7 cm in the base and a height of 16 cm.

Science Applications

Measure in SI

The metric system of measurement was developed in 1795. A modern form of the metric system, called the International System (SI), was adopted in 1960 and provides the standard measurements that all scientists around the world can understand.

The SI system is convenient because unit sizes vary by powers of 10. Prefixes are used to name units. Look at **Table 3** for some common SI prefixes and their meanings.

Table 3 Common SI Prefixes

Prefix	Symbol	Meaning	
kilo-	k	1,000	thousand
hecto-	h	100	hundred
deka-	da	10	ten
deci-	d	0.1	tenth
centi-	c	0.01	hundredth
milli-	m	0.001	thousandth

Example How many grams equal one kilogram?

Step 1 Find the prefix *kilo* in **Table 3.**

Step 2 Using **Table 3,** determine the meaning of *kilo.* According to the table, it means 1,000. When the prefix *kilo* is added to a unit, it means that there are 1,000 of the units in a "*kilo*unit."

Step 3 Apply the prefix to the units in the question. The units in the question are grams. There are 1,000 grams in a kilogram.

Practice Problem Is a milligram larger or smaller than a gram? How many of the smaller units equal one larger unit? What fraction of the larger unit does one smaller unit represent?

Dimensional Analysis

Convert SI Units In science, quantities such as length, mass, and time sometimes are measured using different units. A process called dimensional analysis can be used to change one unit of measure to another. This process involves multiplying your starting quantity and units by one or more conversion factors. A conversion factor is a ratio equal to one and can be made from any two equal quantities with different units. If 1,000 mL equal 1 L then two ratios can be made.

$$\frac{1{,}000\text{ mL}}{1\text{ L}} = \frac{1\text{ L}}{1{,}000\text{ mL}} = 1$$

One can covert between units in the SI system by using the equivalents in **Table 3** to make conversion factors.

Example 1 How many cm are in 4 m?

Step 1 Write conversion factors for the units given. From **Table 3,** you know that 100 cm = 1 m. The conversion factors are

$$\frac{100\text{ cm}}{1\text{ m}} \quad \textit{and} \quad \frac{1\text{ m}}{100\text{ cm}}$$

Step 2 Decide which conversion factor to use. Select the factor that has the units you are converting from (m) in the denominator and the units you are converting to (cm) in the numerator.

$$\frac{100\text{ cm}}{1\text{ m}}$$

Step 3 Multiply the starting quantity and units by the conversion factor. Cancel the starting units with the units in the denominator. There are 400 cm in 4 m.

$$4\,\cancel{\text{m}} \times \frac{100\text{ cm}}{1\,\cancel{\text{m}}} = 400\text{ cm}$$

Practice Problem How many milligrams are in one kilogram? (Hint: You will need to use two conversion factors from **Table 3.**)

Math Skill Handbook

Table 4 Unit System Equivalents

Type of Measurement	Equivalent
Length	1 in = 2.54 cm
	1 yd = 0.91 m
	1 mi = 1.61 km
Mass and Weight*	1 oz = 28.35 g
	1 lb = 0.45 kg
	1 ton (short) = 0.91 tonnes (metric tons)
	1 lb = 4.45 N
Volume	1 in^3 = 16.39 cm^3
	1 qt = 0.95 L
	1 gal = 3.78 L
Area	1 in^2 = 6.45 cm^2
	1 yd^2 = 0.83 m^2
	1 mi^2 = 2.59 km^2
	1 acre = 0.40 hectares
Temperature	$°C = \dfrac{(°F - 32)}{1.8}$
	K = °C + 273

*Weight is measured in standard Earth gravity.

Convert Between Unit Systems Table 4 gives a list of equivalents that can be used to convert between English and SI units.

Example If a meterstick has a length of 100 cm, how long is the meterstick in inches?

Step 1 Write the conversion factors for the units given. From **Table 4,** 1 in = 2.54 cm.

$$\frac{1 \text{ in}}{2.54 \text{ cm}} \quad and \quad \frac{2.54 \text{ cm}}{1 \text{ in}}$$

Step 2 Determine which conversion factor to use. You are converting from cm to in. Use the conversion factor with cm on the bottom.

$$\frac{1 \text{ in}}{2.54 \text{ cm}}$$

Step 3 Multiply the starting quantity and units by the conversion factor. Cancel the starting units with the units in the denominator. Round your answer based on the number of significant figures in the conversion factor.

$$100 \text{ cm} \times \frac{1 \text{ in}}{2.54 \text{ cm}} = 39.37 \text{ in}$$

The meterstick is 39.4 in long.

Practice Problem A book has a mass of 5 lbs. What is the mass of the book in kg?

Practice Problem Use the equivalent for in and cm (1 in = 2.54 cm) to show how 1 in^3 = 16.39 cm^3.

Precision and Significant Digits

When you make a measurement, the value you record depends on the precision of the measuring instrument. This precision is represented by the number of significant digits recorded in the measurement. When counting the number of significant digits, all digits are counted except zeros at the end of a number with no decimal point such as 2,050, and zeros at the beginning of a decimal such as 0.03020. When adding or subtracting numbers with different precision, round the answer to the smallest number of decimal places of any number in the sum or difference. When multiplying or dividing, the answer is rounded to the smallest number of significant digits of any number being multiplied or divided.

Example The lengths 5.28 and 5.2 are measured in meters. Find the sum of these lengths and record your answer using the correct number of significant digits.

Step 1 Find the sum.

5.28 m	2 digits after the decimal
+ 5.2 m	1 digit after the decimal
10.48 m	

Step 2 Round to one digit after the decimal because the least number of digits after the decimal of the numbers being added is 1.

The sum is 10.5 m.

Practice Problem How many significant digits are in the measurement 7,071,301 m? How many significant digits are in the measurement 0.003010 g?

Practice Problem Multiply 5.28 and 5.2 using the rule for multiplying and dividing. Record the answer using the correct number of significant digits.

Scientific Notation

Many times numbers used in science are very small or very large. Because these numbers are difficult to work with scientists use scientific notation. To write numbers in scientific notation, move the decimal point until only one non-zero digit remains on the left. Then count the number of places you moved the decimal point and use that number as a power of ten. For example, the average distance from the Sun to Mars is 227,800,000,000 m. In scientific notation, this distance is 2.278×10^{11} m. Because you moved the decimal point to the left, the number is a positive power of ten.

The mass of an electron is about 0.000 000 000 000 000 000 000 000 000 000 911 kg. Expressed in scientific notation, this mass is 9.11×10^{-31} kg. Because the decimal point was moved to the right, the number is a negative power of ten.

Example Earth is 149,600,000 km from the Sun. Express this in scientific notation.

Step 1 Move the decimal point until one non-zero digit remains on the left.
1.496 000 00

Step 2 Count the number of decimal places you have moved. In this case, eight.

Step 3 Show that number as a power of ten, 10^8.

The Earth is 1.496×10^8 km from the Sun.

Practice Problem How many significant digits are in 149,600,000 km? How many significant digits are in 1.496×10^8 km?

Practice Problem Parts used in a high performance car must be measured to 7×10^{-6} m. Express this number as a decimal.

Practice Problem A CD is spinning at 539 revolutions per minute. Express this number in scientific notation.

Make and Use Graphs

Data in tables can be displayed in a graph—a visual representation of data. Common graph types include line graphs, bar graphs, and circle graphs.

Line Graph A line graph shows a relationship between two variables that change continuously. The independent variable is changed and is plotted on the *x*-axis. The dependent variable is observed, and is plotted on the *y*-axis.

Example Draw a line graph of the data below from a cyclist in a long-distance race.

Table 5 Bicycle Race Data	
Time (h)	Distance (km)
0	0
1	8
2	16
3	24
4	32
5	40

Step 1 Determine the *x*-axis and *y*-axis variables. Time varies independently of distance and is plotted on the *x*-axis. Distance is dependent on time and is plotted on the *y*-axis.

Step 2 Determine the scale of each axis. The *x*-axis data ranges from 0 to 5. The *y*-axis data ranges from 0 to 40.

Step 3 Using graph paper, draw and label the axes. Include units in the labels.

Step 4 Draw a point at the intersection of the time value on the *x*-axis and corresponding distance value on the *y*-axis. Connect the points and label the graph with a title, as shown in **Figure 20.**

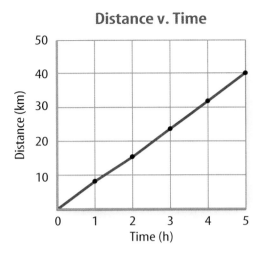

Distance v. Time

Figure 20 This line graph shows the relationship between distance and time during a bicycle ride.

Practice Problem A puppy's shoulder height is measured during the first year of her life. The following measurements were collected: (3 mo, 52 cm), (6 mo, 72 cm), (9 mo, 83 cm), (12 mo, 86 cm). Graph this data.

Find a Slope The slope of a straight line is the ratio of the vertical change, rise, to the horizontal change, run.

$$\text{Slope} = \frac{\text{vertical change (rise)}}{\text{horizontal change (run)}} = \frac{\text{change in } y}{\text{change in } x}$$

Example Find the slope of the graph in **Figure 20.**

Step 1 You know that the slope is the change in *y* divided by the change in *x*.
$$\text{Slope} = \frac{\text{change in } y}{\text{change in } x}$$

Step 2 Determine the data points you will be using. For a straight line, choose the two sets of points that are the farthest apart.
$$\text{Slope} = \frac{(40-0) \text{ km}}{(5-0) \text{ hr}}$$

Step 3 Find the change in *y* and *x*.
$$\text{Slope} = \frac{40 \text{ km}}{5 \text{ h}}$$

Step 4 Divide the change in *y* by the change in *x*.
$$\text{Slope} = \frac{8 \text{ km}}{\text{h}}$$

The slope of the graph is 8 km/h.

Bar Graph To compare data that does not change continuously you might choose a bar graph. A bar graph uses bars to show the relationships between variables. The *x*-axis variable is divided into parts. The parts can be numbers such as years, or a category such as a type of animal. The *y*-axis is a number and increases continuously along the axis.

Example A recycling center collects 4.0 kg of aluminum on Monday, 1.0 kg on Wednesday, and 2.0 kg on Friday. Create a bar graph of this data.

Step 1 Select the *x*-axis and *y*-axis variables. The measured numbers (the masses of aluminum) should be placed on the *y*-axis. The variable divided into parts (collection days) is placed on the *x*-axis.

Step 2 Create a graph grid like you would for a line graph. Include labels and units.

Step 3 For each measured number, draw a vertical bar above the *x*-axis value up to the *y*-axis value. For the first data point, draw a vertical bar above Monday up to 4.0 kg.

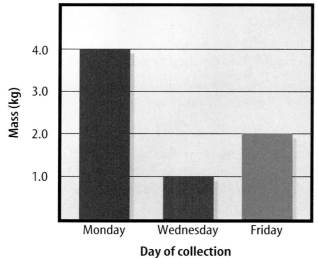

Aluminum Collected During Week

Practice Problem Draw a bar graph of the gases in air: 78% nitrogen, 21% oxygen, 1% other gases.

Circle Graph To display data as parts of a whole, you might use a circle graph. A circle graph is a circle divided into sections that represent the relative size of each piece of data. The entire circle represents 100%, half represents 50%, and so on.

Example Air is made up of 78% nitrogen, 21% oxygen, and 1% other gases. Display the composition of air in a circle graph.

Step 1 Multiply each percent by 360° and divide by 100 to find the angle of each section in the circle.

$$78\% \times \frac{360°}{100} = 280.8°$$

$$21\% \times \frac{360°}{100} = 75.6°$$

$$1\% \times \frac{360°}{100} = 3.6°$$

Step 2 Use a compass to draw a circle and to mark the center of the circle. Draw a straight line from the center to the edge of the circle.

Step 3 Use a protractor and the angles you calculated to divide the circle into parts. Place the center of the protractor over the center of the circle and line the base of the protractor over the straight line.

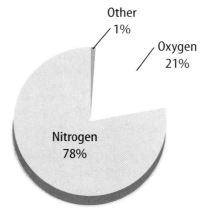

Practice Problem Draw a circle graph to represent the amount of aluminum collected during the week shown in the bar graph to the left.

Weather Map Symbols

Sample Station Model

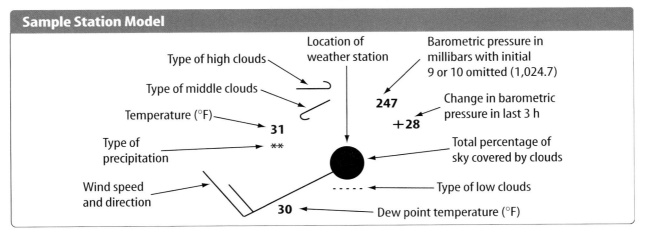

Type of high clouds
Type of middle clouds
Temperature (°F) — **31**
Type of precipitation — **✶✶**
Wind speed and direction
Location of weather station
Barometric pressure in millibars with initial 9 or 10 omitted (1,024.7) — **247**
Change in barometric pressure in last 3 h — **+28**
Total percentage of sky covered by clouds
Type of low clouds — **- - - - -**
Dew point temperature (°F) — **30**

Sample Plotted Report at Each Station

Precipitation	Wind Speed and Direction	Sky Coverage	Some Types of High Clouds
☰ Fog	◯ 0 calm	◯ No cover	Scattered cirrus
★ Snow	1–2 knots	◑ 1/10 or less	Dense cirrus in patches
● Rain	3–7 knots	◕ 2/10 to 3/10	Veil of cirrus covering entire sky
⊼ Thunderstorm	8–12 knots	◕ 4/10	Cirrus not covering entire sky
＇ Drizzle	13–17 knots	◐ —	
▽ Showers	18–22 knots	◕ 6/10	
	23–27 knots	● 7/10	
	48–52 knots	◑ Overcast with openings	
	1 knot = 1.852 km/h	● Completely overcast	

Some Types of Middle Clouds		Some Types of Low Clouds		Fronts and Pressure Systems	
∕	Thin altostratus layer	⌒	Cumulus of fair weather	(H) or High (L) or Low	Center of high- or low-pressure system
∕∕	Thick altostratus layer	⌣	Stratocumulus	▲▲▲▲	Cold front
∕⌣	Thin altostratus in patches	- - - - -	Fractocumulus of bad weather	●●●●	Warm front
∕⌣	Thin altostratus in bands	——	Stratus of fair weather	▲▲●▲	Occluded front
				▲⌢⌢▼	Stationary front

Minerals

Mineral (formula)	Color	Streak	Hardness	Breakage Pattern	Uses and Other Properties
Graphite (C)	black to gray	black to gray	1–1.5	basal cleavage (scales)	pencil lead, lubricants for locks, rods to control some small nuclear reactions, battery poles
Galena (PbS)	gray	gray to black	2.5	cubic cleavage perfect	source of lead, used for pipes, shields for X rays, fishing equipment sinkers
Hematite (Fe_2O_3)	black or reddish-brown	reddish-brown	5.5–6.5	irregular fracture	source of iron; converted to pig iron, made into steel
Magnetite (Fe_3O_4)	black	black	6	conchoidal fracture	source of iron, attracts a magnet
Pyrite (FeS_2)	light, brassy, yellow	greenish-black	6–6.5	uneven fracture	fool's gold
Talc ($Mg_3 Si_4O_{10}(OH)_2$)	white, greenish	white	1	cleavage in one direction	used for talcum powder, sculptures, paper, and tabletops
Gypsum ($CaSO_4 \cdot 2H_2O$)	colorless, gray, white, brown	white	2	basal cleavage	used in plaster of paris and dry wall for building construction
Sphalerite (ZnS)	brown, reddish-brown, greenish	light to dark brown	3.5–4	cleavage in six directions	main ore of zinc; used in paints, dyes, and medicine
Muscovite ($KAl_3Si_3O_{10}(OH)_2$)	white, light gray, yellow, rose, green	colorless	2–2.5	basal cleavage	occurs in large, flexible plates; used as an insulator in electrical equipment, lubricant
Biotite ($K(Mg,Fe)_3(AlSi_3O_{10})(OH)_2$)	black to dark brown	colorless	2.5–3	basal cleavage	occurs in large, flexible plates
Halite (NaCl)	colorless, red, white, blue	colorless	2.5	cubic cleavage	salt; soluble in water; a preservative

Minerals

Mineral (formula)	Color	Streak	Hardness	Breakage Pattern	Uses and Other Properties
Calcite ($CaCO_3$)	colorless, white, pale blue	colorless, white	3	cleavage in three directions	fizzes when HCl is added; used in cements and other building materials
Dolomite ($CaMg(CO_3)_2$)	colorless, white, pink, green, gray, black	white	3.5–4	cleavage in three directions	concrete and cement; used as an ornamental building stone
Fluorite (CaF_2)	colorless, white, blue, green, red, yellow, purple	colorless	4	cleavage in four directions	used in the manufacture of optical equipment; glows under ultraviolet light
Hornblende $(CaNa)_{2-3}$ $(Mg,Al,$ $Fe)_5-(Al,Si)_2$ Si_6O_{22} $(OH)_2)$	green to black	gray to white	5–6	cleavage in two directions	will transmit light on thin edges; 6-sided cross section
Feldspar ($KAlSi_3O_8$) ($NaAl$ Si_3O_8), ($CaAl_2Si_2$ O_8)	colorless, white to gray, green	colorless	6	two cleavage planes meet at 90° angle	used in the manufacture of ceramics
Augite $((Ca,Na)$ (Mg,Fe,Al) $(Al,Si)_2 O_6)$	black	colorless	6	cleavage in two directions	square or 8-sided cross section
Olivine $((Mg,Fe)_2$ $SiO_4)$	olive, green	none	6.5–7	conchoidal fracture	gemstones, refractory sand
Quartz (SiO_2)	colorless, various colors	none	7	conchoidal fracture	used in glass manufacture, electronic equipment, radios, computers, watches, gemstones

Rocks

Rocks		
Rock Type	**Rock Name**	**Characteristics**
Igneous (intrusive)	Granite	Large mineral grains of quartz, feldspar, hornblende, and mica. Usually light in color.
	Diorite	Large mineral grains of feldspar, hornblende, and mica. Less quartz than granite. Intermediate in color.
	Gabbro	Large mineral grains of feldspar, augite, and olivine. No quartz. Dark in color.
Igneous (extrusive)	Rhyolite	Small mineral grains of quartz, feldspar, hornblende, and mica, or no visible grains. Light in color.
	Andesite	Small mineral grains of feldspar, hornblende, and mica or no visible grains. Intermediate in color.
	Basalt	Small mineral grains of feldspar, augite, and possibly olivine or no visible grains. No quartz. Dark in color.
	Obsidian	Glassy texture. No visible grains. Volcanic glass. Fracture looks like broken glass.
	Pumice	Frothy texture. Floats in water. Usually light in color.
Sedimentary (detrital)	Conglomerate	Coarse grained. Gravel or pebble-size grains.
	Sandstone	Sand-sized grains 1/16 to 2 mm.
	Siltstone	Grains are smaller than sand but larger than clay.
	Shale	Smallest grains. Often dark in color. Usually platy.
Sedimentary (chemical or organic)	Limestone	Major mineral is calcite. Usually forms in oceans and lakes. Often contains fossils.
	Coal	Forms in swampy areas. Compacted layers of organic material, mainly plant remains.
Sedimentary (chemical)	Rock Salt	Commonly forms by the evaporation of seawater.
Metamorphic (foliated)	Gneiss	Banding due to alternate layers of different minerals, of different colors. Parent rock often is granite.
	Schist	Parallel arrangement of sheetlike minerals, mainly micas. Forms from different parent rocks.
	Phyllite	Shiny or silky appearance. May look wrinkled. Common parent rocks are shale and slate.
	Slate	Harder, denser, and shinier than shale. Common parent rock is shale.
Metamorphic (nonfoliated)	Marble	Calcite or dolomite. Common parent rock is limestone.
	Soapstone	Mainly of talc. Soft with greasy feel.
	Quartzite	Hard with interlocking quartz crystals. Common parent rock is sandstone.

Topographic Map Symbols

Topographic Map Symbols

————————	Primary highway, hard surface	～～～～	Index contour
▬▬ ▬ ▬ ▬	Secondary highway, hard surface	··········	Supplementary contour
═══════	Light-duty road, hard or improved surface	～～～	Intermediate contour
==========	Unimproved road	◯	Depression contours
+–+–+–+	Railroad: single track		
+—+—+	Railroad: multiple track	— – – —	Boundaries: national
+++++++	Railroads in juxtaposition	— – – —	State
		— · — – · ·	County, parish, municipal
▪▐ ▨ ▓	Buildings	— — — —	Civil township, precinct, town, barrio
♦♦ [+] cem	Schools, church, and cemetery	— — — –	Incorporated city, village, town, hamlet
▫▱ ▨▨	Buildings (barn, warehouse, etc.)	· — · — · ·	Reservation, national or state
∘ ∘	Wells other than water (labeled as to type)	----------	Small park, cemetery, airport, etc.
●●● ⊘	Tanks: oil, water, etc. (labeled only if water)	— ·· — ··	Land grant
⊙ ⚐	Located or landmark object; windmill	————	Township or range line, U.S. land survey
⤬ ×	Open pit, mine, or quarry; prospect	— — — —	Township or range line, approximate location
	Marsh (swamp)		
	Wooded marsh	～～～	Perennial streams
	Woods or brushwood	— — ←	Elevated aqueduct
	Vineyard	∘ ～	Water well and spring
	Land subject to controlled inundation	～～	Small rapids
	Submerged marsh	～～	Large rapids
	Mangrove	◌	Intermittent lake
	Orchard	～～	Intermittent stream
	Scrub	· — · — · ←	Aqueduct tunnel
	Urban area		Glacier
		～～	Small falls
x7369	Spot elevation	～≋	Large falls
670	Water elevation		Dry lake bed

PERIODIC TABLE OF THE ELEMENTS

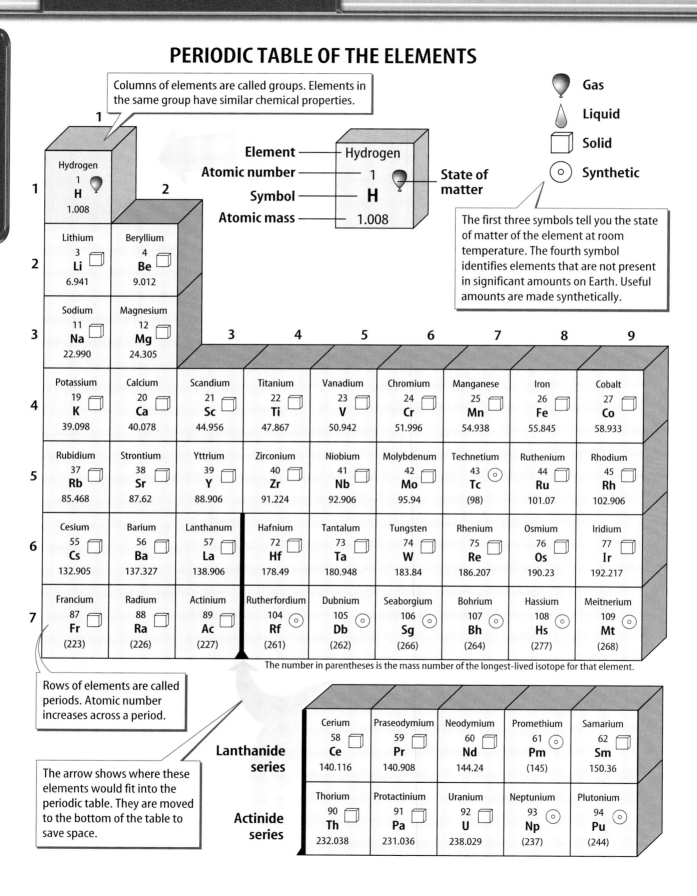

Metal

Metalloid

Nonmetal

The color of an element's block tells you if the element is a metal, nonmetal, or metalloid.

Science Online

Visit booki.msscience.com for updates to the periodic table.

						18
13	14	15	16	17		Helium 2 **He** 4.003
Boron 5 **B** 10.811	Carbon 6 **C** 12.011	Nitrogen 7 **N** 14.007	Oxygen 8 **O** 15.999	Fluorine 9 **F** 18.998		Neon 10 **Ne** 20.180

10	11	12						
			Aluminum 13 **Al** 26.982	Silicon 14 **Si** 28.086	Phosphorus 15 **P** 30.974	Sulfur 16 **S** 32.065	Chlorine 17 **Cl** 35.453	Argon 18 **Ar** 39.948
Nickel 28 **Ni** 58.693	Copper 29 **Cu** 63.546	Zinc 30 **Zn** 65.409	Gallium 31 **Ga** 69.723	Germanium 32 **Ge** 72.64	Arsenic 33 **As** 74.922	Selenium 34 **Se** 78.96	Bromine 35 **Br** 79.904	Krypton 36 **Kr** 83.798
Palladium 46 **Pd** 106.42	Silver 47 **Ag** 107.868	Cadmium 48 **Cd** 112.411	Indium 49 **In** 114.818	Tin 50 **Sn** 118.710	Antimony 51 **Sb** 121.760	Tellurium 52 **Te** 127.60	Iodine 53 **I** 126.904	Xenon 54 **Xe** 131.293
Platinum 78 **Pt** 195.078	Gold 79 **Au** 196.967	Mercury 80 **Hg** 200.59	Thallium 81 **Tl** 204.383	Lead 82 **Pb** 207.2	Bismuth 83 **Bi** 208.980	Polonium 84 **Po** (209)	Astatine 85 **At** (210)	Radon 86 **Rn** (222)
Darmstadtium 110 **Ds** (281)	Unununium * 111 **Uuu** (272)	Ununbium * 112 **Uub** (285)		Ununquadium * 114 **Uuq** (289)		** 116		** 118

* The names and symbols for elements 111–114 are temporary. Final names will be selected when the elements' discoveries are verified.

** Elements 116 and 118 were thought to have been created. The claim was retracted because the experimental results could not be repeated.

Europium 63 **Eu** 151.964	Gadolinium 64 **Gd** 157.25	Terbium 65 **Tb** 158.925	Dysprosium 66 **Dy** 162.500	Holmium 67 **Ho** 164.930	Erbium 68 **Er** 167.259	Thulium 69 **Tm** 168.934	Ytterbium 70 **Yb** 173.04	Lutetium 71 **Lu** 174.967
Americium 95 **Am** (243)	Curium 96 **Cm** (247)	Berkelium 97 **Bk** (247)	Californium 98 **Cf** (251)	Einsteinium 99 **Es** (252)	Fermium 100 **Fm** (257)	Mendelevium 101 **Md** (258)	Nobelium 102 **No** (259)	Lawrencium 103 **Lr** (262)

Glossary/Glosario

Cómo usar el glosario en español:
1. Busca el término en inglés que desees encontrar.
2. El término en español, junto con la definición, se encuentran en la columna de la derecha.

Pronunciation Key

Use the following key to help you sound out words in the glossary.

a................	back (BAK)	ew	food (FEWD)
ay..............	day (DAY)	yoo	pure (PYOOR)
ah.............	father (FAH thur)	yew	few (FYEW)
ow	flower (FLOW ur)	uh	comma (CAH muh)
ar.............	car (CAR)	u (+ con)......	rub (RUB)
e...............	less (LES)	sh.............	shelf (SHELF)
ee.............	leaf (LEEF)	ch.............	nature (NAY chur)
ih.............	trip (TRIHP)	g.............	gift (GIHFT)
i (i + con + e) ..	idea (i DEE uh)	j	gem (JEM)
oh	go (GOH)	ing............	sing (SING)
aw	soft (SAWFT)	zh.............	vision (VIH zhun)
or.............	orbit (OR buht)	k.............	cake (KAYK)
oy..............	coin (COYN)	s	seed, cent (SEED, SENT)
oo	foot (FOOT)	z.............	zone, raise (ZOHN, RAYZ)

English — **A** — **Español**

acid rain: rain, snow, fog, and other forms of precipitation with a pH below 5.6 that can harm plant and animal life and is formed when sulfur dioxide and nitrogen oxides combine with moisture in the atmosphere. (p. 99)

lluvia ácida: lluvia, nieve, neblina y otras formas de precipitación con un pH inferior a 5.6 que puede ser perjudicial para las formas de vida vegetal y animal, y que se forma cuando el dióxido de azufre y los óxidos de nitrógeno se combinan con la humedad de la atmósfera. (p. 99)

adaptation: any structural or behavioral change that helps an organism survive in its particular environment. (p. 70)

adaptación: cualquier cambio de estructura o comportamiento que ayude a un organismo a sobrevivir en su medio ambiente particular. (p. 70)

air mass: large body of air that has the same characteristics of temperature and moisture content as the part of Earth's surface over which it formed. (p. 44)

masa de aire: gran cuerpo de aire que tiene las mismas características de temperatura y contenido de humedad que la parte de la superficie terrestre sobre la cual se formó. (p. 44)

air quality standard: level that a pollutant cannot exceed in ambient air. (p. 112)

estándar de calidad del aire: nivel que un contaminante no puede exceder en el aire ambiental. (p. 112)

ambient (AM bee unt) air: the open, surrounding air that you breathe; any unconfined portion of the atmosphere. (p. 112)

aire ambiental: el aire libre que nos rodea y que respiramos; cualquier porción no confinada de la atmósfera. (p. 112)

atmosphere: Earth's air, which is made up of a thin layer of gases, solids, and liquids; forms a protective layer around the planet and is divided into five distinct layers. (p. 8)

atmósfera: el aire de la Tierra; está compuesta por una capa fina de gases, sólidos y líquidos, forma una capa protectora alrededor del planeta y está dividida en cinco capas distintas. (p. 8)

B

biomagnification (BI oh mag nuh fuh KAY shun): process in which pollutant levels increase through the food chain. (p. 107)

blizzard: winter storm that lasts at least three hours with temperatures of $-12°C$ or below, poor visibility, and winds of at least 51 km/h. (p. 51)

biomagnificación: proceso mediante el cual los niveles de contaminantes aumentan a través de la cadena alimenticia. (p. 107)

nevasca: tormenta invernal que dura por lo menos tres horas con temperaturas de $-12°C$ o menores, escasa visibilidad y vientos de por lo menos 51 km/h. (p. 51)

C

cataract: clouding of the eye's lens, linked to exposure to high amounts of ultraviolet radiation. (p. 106)

chlorofluorocarbons (CFCs): group of chemical compounds used in refrigerators, air conditioners, foam packaging, and aerosol sprays that may enter the atmosphere and destroy ozone. (p. 14)

climate: average weather pattern in an area over a long period of time; can be classified by temperature, humidity, precipitation, and vegetation. (p. 66)

condensation: process in which water vapor changes to a liquid. (p. 19)

conduction: transfer of energy that occurs when molecules bump into each other. (p. 18)

convection: transfer of heat by the flow of material. (p. 18)

Coriolis (kor ee OH lus) effect: causes moving air and water to turn left in the southern hemisphere and turn right in the northern hemisphere due to Earth's rotation. (p. 22)

catarata: nubosidad del lente ocular asociada con la excesiva exposición a la radiación ultravioleta. (p. 106)

clorofluorocarbonos (CFCs): grupo de compuestos químicos usados en refrigeradores, acondicionadores de aire, espumas de empaque y aerosoles; pueden entrar en la atmósfera y destruir el ozono. (p. 14)

clima: modelo meteorológico en un área durante un periodo de tiempo largo; puede clasificarse por temperatura, humedad, precipitación y vegetación. (p. 66)

condensación: proceso mediante el cual el vapor de agua cambia a su forma líquida. (p. 19)

conducción: transferencia de energía que ocurre cuando las moléculas chocan unas con otras. (p. 18)

convección: transferencia de calor mediante flujo de material. (p. 18)

efecto de Coriolis: causa el movimiento del aire y agua hacia la izquierda en el hemisferio sur y hacia la derecha en el hemisferio norte; este efecto es debido a la rotación de la Tierra. (p. 22)

D

deforestation: destruction and cutting down of forests—often to clear land for mining, roads, and grazing of cattle—resulting in increased atmospheric CO_2 levels. (p. 83)

dew point: temperature at which air is saturated and condensation forms. (p. 39)

El Niño (el NEEN yoh): climatic event that begins in the tropical Pacific Ocean; may occur when trade winds weaken or reverse, and can disrupt normal temperature and precipitation patterns around the world. (p. 75)

deforestación: destrucción y tala de los bosques—a menudo el despeje de la tierra para minería, carreteras y ganadería—resultando en el aumento de los niveles atmosféricos de dióxido de carbono. (p. 83)

punto de condensación: temperatura a la que el aire se satura y se genera la condensación. (p. 39)

El Niño: evento climático que comienza en el Océano Pacífico tropical; puede ocurrir cuando los vientos alisios se debilitan o se invierten; puede desestabilizar los patrones normales de precipitación y temperatura del mundo. (p. 75)

Glossary/Glosario

E

emission (ee MIH shun): pollutant released into the air from a given source, such as an automobile tailpipe. (p. 112)

emisión: contaminante expulsado al aire mediante una fuente determinada, tal como el tubo de escape de un automóvil. (p. 112)

F

fog: a stratus cloud that forms when air is cooled to its dew point near the ground. (p. 41)

front: boundary between two air masses with different temperatures, density, or moisture; can be cold, warm, occluded, and stationary. (p. 45)

niebla: nube de estrato que se forma cuando el aire se enfría a su punto de condensación cerca del suelo. (p. 41)

frente: límite entre dos masas de aire con temperatura, densidad o humedad diferentes; puede ser frío, caliente, ocluido o estacionario. (p. 45)

G

global warming: increase in the average global temperature of Earth. (p. 82)

greenhouse effect: natural heating that occurs when certain gases in Earth's atmosphere, such as methane, CO_2, and water vapor, trap heat. (p. 81)

calentamiento global: incremento del promedio de la temperatura global. (p. 82)

efecto invernadero: calentamiento natural que ocurre cuando ciertos gases en la atmósfera terrestre, como el metano, el dióxido de carbono y el vapor de agua atrapan el calor. (p. 81)

H

hibernation: behavioral adaptation for winter survival in which an animal's activity is greatly reduced, its body temperature drops, and body processes slow down. (p. 72)

humidity: amount of water vapor held in the air. (p. 38)

hurricane: large, severe storm that forms over tropical oceans, has winds of at least 120 km/h, and loses power when it reaches land. (p. 50)

hydrosphere: all the waters of Earth. (p. 19)

hibernación: adaptación del comportamiento para sobrevivir durante el invierno en la cual la actividad del animal se ve fuertemente reducida, su temperatura corporal se reduce y los procesos corporales disminuyen su ritmo. (p. 72)

humedad: cantidad de vapor de agua suspendido en el aire. (p. 38)

huracán: tormenta grande y severa que se forma sobre los océanos tropicales, tiene vientos de por lo menos 120 km/h y pierde su fuerza cuando alcanza la costa. (p. 50)

hidrosfera: toda el agua de la Tierra. (p. 19)

I

ionosphere: layer of electrically charged particles in the thermosphere that absorbs AM radio waves during the day and reflects them back at night. (p. 11)

ionosfera: capa de partículas con carga eléctrica presentes en la termosfera, la cual absorbe las ondas de radio AM durante el día y las refleja durante la noche. (p. 11)

isobars: lines drawn on a weather map that connect points having equal atmospheric pressure; also indicate the location of high- and low-pressure areas and can show wind speed. (p. 53)

isobaras: líneas dibujadas en un mapa meteorológico que conectan los puntos que tienen una presión atmosférica similar; también indican la ubicación de las áreas de baja y alta presión y pueden mostrar la velocidad del viento. (p. 53)

isotherm (I suh thurm): line drawn on a weather map that connects points having equal temperature. (p. 53)

isoterma: línea dibujada en un mapa meteorológico que conecta los puntos que tienen la misma temperatura. (p. 53)

jet stream: narrow belt of strong winds that blows near the top of the troposphere. (p. 24)

corriente de chorro: faja angosta de vientos fuertes que soplan cerca de la parte superior de la troposfera. (p. 24)

L

land breeze: movement of air from land to sea at night, created when cooler, denser air from the land forces up warmer air over the sea. (p. 25)

brisa terrestre: movimiento de aire nocturno de la tierra al mar, generado cuando el aire denso y frío proveniente de la tierra empuja hacia arriba al aire caliente que está sobre el mar. (p. 25)

M

meteorologist (meet ee uh RAHL uh just): studies weather and uses information from Doppler radar, weather satellites, computers, and other instruments to make weather maps and provide forecasts. (p. 52)

meteorólogo: persona que estudia el clima y usa información del radar Doppler, satélites meteorológicos, computadoras y otros instrumentos para elaborar mapas del estado del tiempo y hacer pronósticos. (p. 52)

O

ozone layer: layer of the stratosphere with a high concentration of ozone; absorbs most of the Sun's harmful ultraviolet radiation. (pp. 14, 102)

capa de ozono: capa de la estratosfera con una concentración alta de ozono y que absorbe la mayor parte de la radiación ultravioleta dañina del sol. (pp. 14, 102)

P

particulate matter: solid particles and liquid droplets suspended in the air. (p. 100)

materia particulada: partículas sólidas y gotas líquidas suspendidas en el aire. (p. 100)

photochemical smog: brown haze formed when secondary pollutants interact with sunlight. (p. 97)

smog fotoquímico: bruma color café formada cuando los contaminantes secundarios interactúan con la luz solar. (p. 97)

polar zones: climate zones that receive solar radiation at a low angle, extend from 66°N and S latitude to the poles, and are never warm. (p. 66)

zonas polares: zonas climáticas que reciben radiación solar a un ángulo reducido, se extienden desde los 66° de latitud norte y sur hasta los polos y nunca son cálidas. (p. 66)

Glossary/Glosario

precipitation: water falling from clouds—including rain, snow, sleet, and hail—whose form is determined by air temperature. (p. 42)

primary pollutant: substance released directly into the air in a harmful form, including volcanic gases, soot from trucks, and smoke from forest fires. (p. 96)

precipitación: agua que cae de las nubes—incluyendo lluvia, nieve, aguanieve y granizo—cuya forma está determinada por la temperatura del aire. (p. 42)

contaminante primario: sustancia liberada directamente al aire de forma nociuaa, incluyendo los gases volcánicos, el hollín de los camiones el y humo de los incendios forestales. (p. 96)

R

radiation: energy transferred by waves or rays. (p. 18)

relative humidity: measure of the amount of moisture held in the air compared with the amount it can hold at a given temperature; can range from 0 percent to 100 percent. (p. 38)

radiación: energía transmitida por ondas o rayos. (p. 18)

humedad relativa: medida de la cantidad de humedad suspendida en el aire en comparación con la cantidad que puede contener a una temperatura determinada; puede variar del cero al cien por ciento. (p. 38)

S

sea breeze: movement of air from sea to land during the day when cooler air from above the water moves over the land, forcing the heated, less dense air above the land to rise. (p. 25)

season: short period of climate change in an area caused by the tilt of Earth's axis as Earth revolves around the Sun. (p. 74)

secondary pollutant: substance that pollutes the air after reacting with other substances in Earth's atmosphere. (p. 96)

station model: indicates weather conditions at a specific location, using a combination of symbols on a map. (p. 53)

brisa marina: movimiento de aire del mar a la tierra durante el día, cuando el aire frío que está sobre el mar empuja al aire caliente y menos denso que está sobre la tierra. (p. 25)

estación: periodo corto de cambio climático en un área, causado por la inclinación del eje de la Tierra conforme gira alrededor del sol. (p. 74)

contaminante secundario: sustancia que contamina el aire después de reaccionar con otras sustancias de la atmósfera terrestre. (p. 96)

modelo estacional: indica las condiciones del estado del tiempo en una ubicación específica, utilizando una combinación de símbolos en un mapa. (p. 53)

T

temperate zones: climate zones with moderate temperatures that are located between the tropics and the polar zones. (p. 66)

tornado: violent, whirling windstorm that crosses land in a narrow path and can result from wind shears inside a thunderhead. (p. 48)

toxic air pollutant: substance released into the air that can cause health problems, including cancer. (p. 101)

tropics: climate zone that receives the most solar radiation, is located between latitudes 23°N and 23°S, and is always hot, except at high elevations. (p. 66)

zonas templadas: zonas climáticas con temperaturas moderadas que están localizadas entre los trópicos y las zonas polares. (p. 66)

tornado: tormenta de viento en forma de remolino que cruza la tierra en un curso estrecho y puede resultar de vientos que se entrecruzan en direcciones opuestas dentro del frente de una tormenta. (p. 48)

contaminante atmosférico tóxico: sustancia expulsada al aire y que puede causar problemas de salud como el cáncer. (p. 101)

trópicos: zonas climáticas que reciben la mayor parte de la radiación solar, están localizadas entre los 23° de latitud norte y 23° de latitud sur y siempre son cálidas excepto a grandes alturas. (p. 66)

troposphere/weather | **troposfera/estado del tiempo**

troposphere: layer of Earth's atmosphere that is closest to the ground, contains 99 percent of the water vapor and 75 percent of the atmospheric gases, and is where clouds and weather occur. (p. 10)

troposfera: capa de la atmósfera terrestre que se encuentra cerca del suelo, contiene el 99 por ciento del vapor de agua y el 75 por ciento de los gases atmosféricos; es donde se forman las nubes y las condiciones meteorológicas. (p. 10)

ultraviolet radiation: a type of energy that comes to Earth from the Sun, can damage skin and cause cancer, and is mostly absorbed by the ozone layer. (pp. 14, 106)

radiación ultravioleta: tipo de energía que llega a la Tierra desde el sol y que puede dañar la piel y causar cáncer; la mayor parte de esta radiación es absorbida por la capa de ozono. (pp. 14, 106)

weather: state of the atmosphere at a specific time and place, determined by factors including air pressure, amount of moisture in the air, temperature, wind, and precipitation. (p. 36)

estado del tiempo: estado de la atmósfera en un momento y lugar específicos, determinado por factores que incluyen la presión del aire, cantidad de humedad en el aire, temperatura, viento y precipitación. (p. 36)

Index

Italic numbers = illustration/photo   **Bold numbers** = **vocabulary term**
lab = indicates a page on which the entry is used in a lab
act = indicates a page on which the entry is used in an activity

Index

Index

Credits

Magnification Key: Magnifications listed are the magnifications at which images were originally photographed.
LM–Light Microscope
SEM–Scanning Electron Microscope
TEM–Transmission Electron Microscope

Acknowledgments: Glencoe would like to acknowledge the artists and agencies who participated in illustrating this program: Absolute Science Illustration; Andrew Evansen; Argosy; Articulate Graphics; Craig Attebery represented by Frank & Jeff Lavaty; CHK America; John Edwards and Associates; Gagliano Graphics; Pedro Julio Gonzalez represented by Melissa Turk & The Artist Network; Robert Hynes represented by Mendola Ltd.; Morgan Cain & Associates; JTH Illustration; Laurie O'Keefe; Matthew Pippin represented by Beranbaum Artist's Representative; Precision Graphics; Publisher's Art; Rolin Graphics, Inc.; Wendy Smith represented by Melissa Turk & The Artist Network; Kevin Torline represented by Berendsen and Associates, Inc.; WILDlife ART; Phil Wilson represented by Cliff Knecht Artist Representative; Zoo Botanica.

Photo Credits

Cover NASA/Science Photo Library/Photo Researchers; **i ii** NASA/Science Photo Library/Photo Researchers; **iv** (bkgd)John Evans, (inset)NASA/Science Photo Library/ Photo Researchers; **v** (t)PhotoDisc, (b)John Evans; **vi** (l)John Evans, (r)Geoff Butler; **vii** (l)John Evans, (r)PhotoDisc; **viii** PhotoDisc; **ix** Aaron Haupt Photography; **x** Roy Johnson/ Tom Stack & Assoc.; **xi** Holger Weitzel/CORBIS; **xii** Roy Morsch/The Stock Market/CORBIS; **1** David Weintraub/Stock Boston; **2** (t)Galen Rowell/Corbis, (b)Jeffrey Howe/Visuals Unlimited; **3** (t)Poav Levy/PhotoTake NYC/PictureQues, (b)AP/Wide World Photos/Dave Martin; **4** Lawrence Migdale/ Stock Boston/PictureQuest; **5** Luis M. Alvarez/AP/Wide World Photos; **6–7** S.P. Gillette/CORBIS; **8** NASA; **9** (l)Frank Rossotto/The Stock Market/CORBIS, (r)Larry Lee/CORBIS; **12** Laurence Fordyce/CORBIS; **14** Doug Martin; **15** NASA/ GSFC; **16** Michael Newman/PhotoEdit, Inc.; **21** (t)Dan Guravich/Photo Researchers, (b)Bill Brooks/Masterfile; **23** (cw from top)Gene Moore/PhotoTake NYC/PictureQuest, Phil Schermeister/CORBIS, Joel W. Rogers, Kevin Schafer/ CORBIS, Stephen R. Wagner; **24** Bill Brooks/Masterfile; **26 27** David Young-Wolff/PhotoEdit, Inc.; **28** Bob Rowan/ CORBIS; **34–35** Reuters NewMedia, Inc./CORBIS; **35** KS Studios; **36** Kevin Horgan/Stone/Getty Images; **37** Fabio Colombini/Earth Scenes; **41** (t)Charles O'Rear/CORBIS, (b)Joyce Photographics/Photo Researchers; **42** (l)Roy Morsch/The Stock Market/CORBIS, (r)Mark McDermott; **43** (l)Mark E. Gibson/Visuals Unlimited, (r)EPI Nancy Adams/Tom Stack & Assoc.; **45** Van Bucher/Science Source/Photo Researchers; **47** Jeffrey Howe/Visuals Unlimited; **48** Roy Johnson/Tom Stack & Assoc.; **49** (l)Warren Faidley/Weatherstock, (r)Robert Hynes; **50** NASA/Science Photo Library/Photo Researchers; **51** Fritz Pölking/Peter Arnold, Inc.; **58** (bkgd)Erik Rank/Photonica, (others)courtesy Weather Modification Inc.; **59** (l)George D. Lepp/Photo Researchers, (r)Janet Foster/Masterfile; **60** Ruth Dixon; **64–65** Andrew Wenzel/Masterfile; **69** (l)William Leonard/DRK Photo, (r)Bob Rowan, Progressive Image/ CORBIS; **70** John Shaw/Tom Stack & Assoc.; **71** (tl)David Hosking/CORBIS, (tr)Yva Momatiuk & John Eastcott/Photo Researchers, (b)Michael Melford/The Image Bank/Getty Images; **72** (t)S.R. Maglione/Photo Researchers, (c)Jack Grove/Tom Stack & Assoc., (b)Fritz Pölking/Visuals Unlimited; **73** Zig Leszczynski/Animals Animals; **75** (l)Jonathan Head/AP/Wide World Photos, (r)Jim Corwin/Index Stock; **77** (t)A. Ramey/PhotoEdit, Inc., (b)Peter Beck/Pictor; **78** Galen Rowell/Mountain Light; **82** John Bolzan; **83** Chip & Jill Isenhart/Tom Stack & Assoc.; **85** Matt Meadows; **87** Doug Martin; **88** Alberto Garcia/Saba; **89** Steve Kaufman/DRK Photo; **94–95** The Image Bank/Getty Images; **96** (l)Krafft-Explorer/Photo Researchers, (r)Holger Weitzel/CORBIS; **100** (t)Andrew Syred/Science Photo Library/Photo Researchers, (c)David Scharf/Peter Arnold, Inc., (b)Bruce Iverson; **101** (tr)Rob Garbarini/Index Stock, (l)Powerstock-ZEFA/Index Stock, (br)Tony Freeman/ PhotoEdit, Inc.; **104 105** David Young-Wolff/PhotoEdit, Inc.; **106** (l)Dr. P. Marazzi/Photo Researchers, (r)Rafael Macia/ Photo Researchers; **108** William Johnson/Stock Boston; **110** Roger Ressmeyer/CORBIS; **112** Dan Habib/Impact Visuals/PictureQuest; **113** (t)John Sohlden/Visuals Unlimited, (b)Charles D. Winter/Photo Researchers; **114** (t)David Weintraub/Stock Boston, (b)Karl Lugmaier, Viennaslide Photoagency/CORBIS; **117** (t)Oliver Benn/ Stone/Getty Images, (b)David Weintraub/Stock Boston; **118** Bettmann/CORBIS; **119** (l)Nik Wheeler/CORBIS, (r)Owen Franken/Stock Boston/PictureQuest; **124** PhotoDisc; **126** Tom Pantages; **130** Michell D. Bridwell/ PhotoEdit, Inc.; **131** (t)Mark Burnett, (b)Dominic Oldershaw; **132** StudiOhio; **133** Timothy Fuller; **134** Aaron Haupt; **136** KS Studios; **137** Matt Meadows; **138** (t)Matt Meadows, (b)Mary Lou Uttermohlen; **139** Doug Martin; **140** Amanita Pictures; **141** Bob Daemmrich; **143** Davis Barber/PhotoEdit, Inc.; **159** Matt Meadows; **160** (l)Dr. Richard Kessel, (c)NIBSC/Science Photo Library/Photo Researchers, (r)David John/Visuals Unlimited; **161** (t)Runk/ Schoenberger from Grant Heilman, (bl)Andrew Syred/ Science Photo Library/Photo Researchers, (br)Rich Brommer; **162** (tr)G.R. Roberts, (l)Ralph Reinhold/Earth Scenes, (br)Scott Johnson/Animals Animals; **163** Martin Harvey/DRK Photo.